SYSTEMS ENGINEERING METHODS, DEVELOPMENTS AND TECHNOLOGY

NETWORKED CONTROL SYSTEMS

THEORY, APPLICATIONS AND ANALYSIS

SYSTEMS ENGINEERING METHODS, DEVELOPMENTS AND TECHNOLOGY

Additional books and e-books in this series can be found
on Nova's website under the Series tab.

Systems Engineering Methods, Developments and Technology

Networked Control Systems

Theory, Applications and Analysis

Shiwen Tong
and
Dianwei Qian
Editors

Copyright © 2021 by Nova Science Publishers, Inc.

All rights reserved. No part of this book may be reproduced, stored in a retrieval system or transmitted in any form or by any means: electronic, electrostatic, magnetic, tape, mechanical photocopying, recording or otherwise without the written permission of the Publisher.

We have partnered with Copyright Clearance Center to make it easy for you to obtain permissions to reuse content from this publication. Simply navigate to this publication's page on Nova's website and locate the "Get Permission" button below the title description. This button is linked directly to the title's permission page on copyright.com. Alternatively, you can visit copyright.com and search by title, ISBN, or ISSN.

For further questions about using the service on copyright.com, please contact:
Copyright Clearance Center
Phone: +1-(978) 750-8400   Fax: +1-(978) 750-4470   E-mail: info@copyright.com.

NOTICE TO THE READER

The Publisher has taken reasonable care in the preparation of this book, but makes no expressed or implied warranty of any kind and assumes no responsibility for any errors or omissions. No liability is assumed for incidental or consequential damages in connection with or arising out of information contained in this book. The Publisher shall not be liable for any special, consequential, or exemplary damages resulting, in whole or in part, from the readers' use of, or reliance upon, this material. Any parts of this book based on government reports are so indicated and copyright is claimed for those parts to the extent applicable to compilations of such works.

Independent verification should be sought for any data, advice or recommendations contained in this book. In addition, no responsibility is assumed by the Publisher for any injury and/or damage to persons or property arising from any methods, products, instructions, ideas or otherwise contained in this publication.

This publication is designed to provide accurate and authoritative information with regard to the subject matter covered herein. It is sold with the clear understanding that the Publisher is not engaged in rendering legal or any other professional services. If legal or any other expert assistance is required, the services of a competent person should be sought. FROM A DECLARATION OF PARTICIPANTS JOINTLY ADOPTED BY A COMMITTEE OF THE AMERICAN BAR ASSOCIATION AND A COMMITTEE OF PUBLISHERS.

Additional color graphics may be available in the e-book version of this book.

Library of Congress Cataloging-in-Publication Data

ISBN: 978-1-53619-892-8

Published by Nova Science Publishers, Inc. † New York

CONTENTS

Preface   vii

Chapter 1   Linear-Model-Predictor-Based Networked Predictive Fuzzy Control of Systems with Forward Channel Delays   1
Shiwen Tong and Dianwei Qian

Chapter 2   Design of a Data-Based Networked Tracking Control System   19
Shiwen Tong and Dianwei Qian

Chapter 3   Design and Implementation of a Data-Based Adaptative Networked Tracking Control System with NetCon   41
Shiwen Tong and Ye Zhao

Chapter 4   PID Temperature Control for an Air Tank System with Parameters Tuning Through Network   59
Shiwen Tong and Dianwei Qian

Chapter 5   On Zeno Behavior in Event-Triggered Control of Networked Systems   73
Hao Yu, Fei Hao and Tongwen Chen

Chapter 6	Robust Guaranteed Performance Consensus for MAS with Time-Delays and Uncertainty *Yaxiao Zhang and Shiwen Tong*	**115**
Chapter 7	Robust Guaranteed Performance Formation Control for MAS with Uncertain Topologies *Yaxiao Zhang and Shiwen Tong*	**133**
About the Editors		**153**
Index		**155**

PREFACE

Networked Control System (NCS) can be regarded as one of the special types of control system in which sensors, controllers and actuators are connected to a closed-loop. Due to the media-sharing characteristic, time-delay, data packet dropout and data displacement etc. are inevitable phenomenon in such a control system, which can greatly degrade the control performance, even make the control system unstable. Alleviating the effects of them have become the most attractive research hotspots in the last two decades. In fact, all the above three problems can be summarized as the time-delay issue. There are two kinds of time-delay compensation strategy. One is active compensation. The other is passive compensation. For the former, prediction is the core idea. Selecting the appropriate candidate predicted control action according to the time delay information is a feasible solution. For the latter, making the system insensitive to delay is a good choice. The book covers the design, modeling, control, simulation and application of the networked control system.

Apparently, the book cannot include all research topics. The editor and the authors wish that it could reveal some tendencies on this research field and benefit readers. In this book, different aspects of networked control are explored. Chapters includes some new tendencies and developments in research on Networked Predictive Fuzzy Control (NPFC), Data-based Networked Tracking Control (DNTC), Data-based Adaptive Networked Tracking Control (DNATC),

Controller Parameters Tuning through Network, Event-Triggered Control of Networked System and Multi-Agent Systems (MAS).

Chapter 1 – This chapter presents a novel networked control framework, using fuzzy logic control, for systems with fixed and random network delays which are known to greatly weaken the control performance of the controlled system. To deal with the network delays, the predicted differences between the desired future set-points and the predicted outputs from a model predictor are utilized as the inputs of a fuzzy controller, thus a series of future control actions are generated. By selecting the appropriated control sequence in the plant side, the network delays including the fixed delays and the random delays in thr forward communication channel are compensated. The simulative results demonstrate that the proposed method can obviously reduce the effect of network delays, and improve the system dynamic performance.

Chapter 2 – Tracking control is a very challenging problem in the Networked Control System (NCS), especially for the process with blurred mechanism and where only input-output data are available. This chapter has proposed a data-based design approach for the Networked Tracking Control System (NTCS). The method utilizes the input-output data of the controlled process to establish a predictive model with the help of Fuzzy Cluster Modelling (FCM) technology. Then, the deduced error and error change in the future are treated as inputs of a Fuzzy Sliding Mode Controller (FSMC) to obtain a string of future control actions. These candidate control actions in the controller side are delivered to the plant side. Thus, the network induced time delays are compensated by selecting appropriate control action. Simulation outputs prove the validity of the proposed method.

Chapter 3 – The time variation of input signal, time delay and data dropout caused by network and other factors will reduce the performance of networked control, especially the networked tracking control of some processes whose mechanism are unclear, only the input and output data of the system are available. In order to solve such problem, we have proposed an adaptive networked tracking control method based on data and realized it on NetCon system. Firstly, the input-output data are obtained by applying the signal excitation to the controlled system, and then the T-S fuzzy model of the controlled object is established by using the fuzzy clustering technology. Then, fuzzy model is transformed into a fuzzy singleton model to obtain fuzzy predictor which can predict factory input at future time. In order to realize active compensation, the fuzzy singleton model is transformed into an inverse model according to the reversibility condition, which can generate the future control action. The adaptive control strategy and the internal model structure are adopted to eliminate the external disturbance and the system uncertainty. Simulation results in NetCon system show the effectiveness of the proposed method.

Chapter 4 – This chapter presents an application of Networked Control in the air tank temperature system. The system is controlled by Proportion-Integration-Differential (PID) algorithm running in the Siemens S7-200 PLC. Through an OPC server component, controller parameters (proportion, integration, differential) can be remote tuning by Matlab. Thus, complex control algorithm such as fuzzy inference, expert system and genetic optimization can be utilized. The process is supervised by configuration software, for example King Views, located in different geographical areas at the same time.

Chapter 5 – In networked systems, Zeno behavior denotes the phenomenon in which an infinite number of transmissions occur in a finite time interval. This phenomenon is extremely undesirable in event-triggered control, which aims at saving communication resources by relating the transmission scheduling with online information. This chapter studies the existence of Zeno behavior in event-triggered control with error-based triggering conditions. It is shown that Zeno behavior is closely related to some particular states, which make the threshold functions in triggering conditions equal to zero. Three kinds of event-triggered control systems, namely, the systems with relative triggering conditions, the finite-time event-triggered control systems, and the systems with external threshold signals, are investigated in detail. The corresponding necessary or sufficient conditions for Zeno behavior are obtained. Based on these analyses, it is discovered that an event-triggered control system with a linear plant can be internally stable but not input-to-state stable with respect to external disturbances; some conflicts between finite-time stability and event-triggered control are pointed out; and the difference between the concepts of Zeno-freeness and an event-separation property is revealed. Several numerical examples and simulations are provided to illustrate the feasibility of the proposed results.

Chapter 6 – In this chapter, a robust guaranteed performance consensus problem for continuous-time linear high-order multi-agent systems with uncertainties and time-varying delays is studied. Firstly, the robust guaranteed performance consensus problem is transformed into a robust guaranteed performance control problem of an auxiliary uncertain system with time-varying delays by a linear transformation. Secondly, a sufficient condition for the robust guaranteed performance consensus problem is presented in terms of linear matrix inequality techniques by robust guaranteed performance control theory, and an upper bound of the guaranteed performance function is given. Finally, a numerical example is shown to demonstrate the above theoretical results.

Chapter 7 – This chapter investigates a robust guaranteed performance formation problem for a class of continuous-time linear high-order multi-agent systems with uncertain communication topology which is modeled by directed graph. Firstly, the robust guaranteed performance formation problem is transformed into a robust guaranteed performance control problem of an auxiliary uncertain system by a linear transformation. Secondly, a sufficient condition for the robust guaranteed performance formation control problem is presented in terms of linear matrix inequality techniques, and an upper bound of the guaranteed performance function is given. Finally, a numerical example is shown to demonstrate the effectiveness of the theoretical results.

Shiwen Tong
College of Robotics
Beijing Union University
Beijing, China

Dianwei Qian
School of Control and Comupter Engineering
North China Electric Power University
Beijing, China

In: Networked Control Systems
Editors: S. Tong and D. Qian
ISBN: 978-1-53619-892-8
© 2021 Nova Science Publishers, Inc.

Chapter 1

LINEAR-MODEL-PREDICTOR-BASED NETWORKED PREDICTIVE FUZZY CONTROL OF SYSTEMS WITH FORWARD CHANNEL DELAYS

Shiwen Tong[1,3,*] *and Dianwei Qian*[2]

[1]College of Robotics, Beijing Union University, Beijing, China
[2]School of Control and Computer Engineering,
North China Electric Power University, Beijing, China
[3]State Key Laboratory for Management and Control of Complex Systems, Institute of Automation, Chinese Academy of Sciences, Beijing, China

ABSTRACT

This chapter presents a novel networked control framework, using fuzzy logic control, for systems with fixed and random network delays which are known to greatly weaken the control performance of the controlled system. To deal with the network delays, the predicted differences between the desired future set-points and the predicted outputs from a model predictor are utilized

---

* Corresponding Author's E-mail: shiwen.tong@buu.edu.cn.

as the inputs of a fuzzy controller, thus a series of future control actions are generated. By selecting the appropriated control sequence in the plant side, the network delays including the fixed delays and the random delays in the forward communication channel are compensated. The simulative results demonstrate that the proposed method can obviously reduce the effect of network delays, and improve the system dynamic performance.

Keywords: networked control, model prediction, fuzzy control, delay compensation

1. INTRODUCTION

The emergence of the network technology has changed the communication architecture of control systems from traditional point-to-point to current common bus. Sensors, actuators and controllers are connected through network, formed a feedback control system (namely, networked control system). This architecture has injected fresh blood to the classic and modern control theories and also arises higher challenges to the controller design at the same time. On the one hand, the introduction of the network to the control system brings many advantages such as low cost, easy maintenance and high reliability. On the other hand, the unavoidable time delay, data dropout and other complicated phenomenon existing in the network should be considered. In recent years, networked control theory and technology have become an important and hot research area. Scholars from different countries have made a lot of breakthroughs in the networked control [1-8].

Network delay has become one of the most concerned issues in the networked control system. Because the network delays can dramatically degrade the control performance of the systems even makes the systems unstable. From recent published literatures, it can be seen that the treatment of the network delays can be summarized in the following: The first one is to integrate delay information into the controller design by designing a robust controller to decrease the effect of time delay [3, 6-8]; The second one is to estimate delay information in the backward or forward channel by using reason-rule-table, average-value or delay windows (DW) method [9, 10]; The third one is to eliminate the network delays in the return path by using a cascade control structure with P (proportion) control in the inner loop and fuzzy adaptive control in the outer loop [4]; As we all known, one of the obvious characteristics of the networked control system is that the communication networks can transmit a packet of data simultaneously. This feature provides another solution to compensate for network delays in the forward channel [11, 12]. Based on it, Liu et al. [1, 2] proposed networked predictive control (NPC), using the strings of future control actions, to compensate for the forward channel delays. In this chapter, we try to design a networked controller by using fuzzy control theories. Different from networked predictive control method, we separate the model prediction from the controller design. The function of the model predictor is just to produce future predicted outputs. A fuzzy controller is designed to generate a series of future control sequence based on the errors between the desired future outputs and the model predicted outputs. Then the strings of future control actions are packed and sent to the plant side through the communication channel. Thus, the effect of delays in the forward channel is lessened by using a delay compensator in the plant side.

Predictive control and fuzzy control are powerful tools. They have been used in the design of the networked controller [1–5, 7, 8]. Some researchers have connected the prediction to the fuzzy control and have proposed 'predictive fuzzy control' method [13, 14]. They use 'future error' and 'future error change' as inputs of the fuzzy controller to produce the control actions at the current time. To my knowledge, the combination of networked control system and the predictive fuzzy control has not been reported except one paper written by us in 2007 [15]. We call it 'networked predictive fuzzy control' with abbreviation NPFC. The core idea of the NPFC is producing 'future control actions' by fuzzy controller design according to the 'future error' and 'future error change' from a model predictor. Then the network delays can be compensated by choosing the 'future control actions.'

This chapter is organized as follows: The architecture of networked predictive fuzzy control (NPFC) is firstly presented. Secondly, a model predictor based on Diophantine equation is proposed. Thirdly, the fuzzy controller using 'future errors' and 'future error changes' as inputs to derive 'future control actions' is designed. Then the delay compensation mechanism is discussed and the method is implemented in a servo control system. Finally, the conclusions are drawn in Section 4.

2. DESIGN OF NETWORKED CONTROL SYSTEMS

2.1. Structure of Networked Predictive Fuzzy Control Systems

The networked predictive fuzzy control system as shown in Figure 1 mainly consists of three key parts: the model predictor, the fuzzy controller and the delay compensator. The model predictor is used to predict future outputs of the controlled system $y(t|t)$, $y(t+1|t)$, $\cdots$, $y(t+N-1|t)$ according to the delayed output $y(t-1)$ of the controlled system in the backward channel and the control actions $u(t-d-1)$, $u(t-d-2)$, $\cdots$, $u(t-d-n_b)$ in the past. The errors $e(t|t)$, $e(t+1|t)$, $\cdots$, $e(t+N-1|t)$ between the desired future outputs $r(t|t)$, $r(t+1|t)$, $\cdots$, $r(t+N-1|t)$ and the predictive future outputs $y(t|t)$, $y(t+1|t)$, $\cdots$, $y(t+N-1|t)$ of the controlled system can be used to design a fuzzy controller to produce the future control sequences $u(t|t)$, $u(t+1|t)$, $\cdots$, $u(t+N_u-1|t)$. Then the future control sequences are packed and sent to the plant side through network. In the plant side, a delay compensator is used to compensate for the forward network delays by selecting appropriate control sequence.

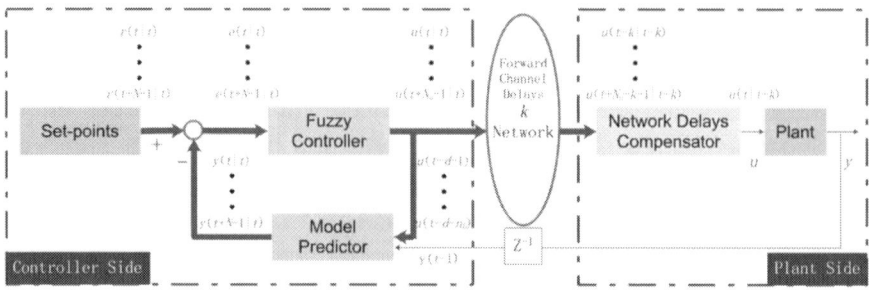

Figure 1. Structure of networked predictive fuzzy control system.

2.2. Model Predictor

The function of the model predictor is to produce a series of future outputs of the controlled object according to the delayed output in the backward channel and the control actions in the past. Models, no matter linear or nonlinear model, of such functions can all be the predictor. To simplified the issue and focus on the research work step by step, we only consider forward channel delays in this chapter. This consideration is reasonable in some cases. For example, in a network that the transmission velocity in the backward channel is faster than the forward channel thus the delays in the backward channel can be neglected. Therefore a linear model predictor based on the Diophantine equation is proposed.

Consider a single-input and single-output process with the following form

$$A(z^{-1})y(t) = z^{-d}B(z^{-1})u(t) \tag{1}$$

where,

$$A(z^{-1}) = 1 + a_1 z^{-1} + \text{L} + a_{n_a} z^{-n_a}$$
$$B(z^{-1}) = b_0 + b_1 z^{-1} + \text{L} + b_{n_b} z^{-n_b}$$

Introducing a Diophantine equation to derive the model predictor.

$$\Delta A(z^{-1})E_i(z^{-1}) + z^{-i}F_i(z^{-1}) = 1 \tag{2}$$

where, $E_i(z^{-1})$ is of order $i - 1$ and $F_i(z^{-1})$ is of order n_a.

$$E_i(z^{-1}) = 1 + \sum_{j=1}^{i-1} e_{i,j} z^{-j}, F_i(z^{-1}) = \sum_{j=0}^{n_a} f_{i,j} z^{-j}$$

Define N as predictive horizon, N_m as model control horizon, from (2.1) and (2.2), the predicted value $Y_p(t + 1)$ of the controlled system can be obtained.

$$Y_d(t+1) = P\Delta U(t-d) + Q\Delta U(t-d-1) + Fy(t-1) \tag{3}$$

where,

$$P = \begin{bmatrix} g_0 & 0 & \cdots & 0 \\ g_1 & g_0 & \cdots & 0 \\ \vdots & \vdots & \ddots & \vdots \\ gN_m-1 & \cdots & \cdots & g_0 \\ \vdots & \vdots & \cdots & \vdots \\ gN-1 & gN-2 & \cdots & gN-N_m \end{bmatrix}$$

$$Y_p(t+1) = [y_p(t\mid t), y_p(t+1\mid t), \cdots, y_p(t+N-1\mid t)]^T$$

$$\Delta U(t-d) = [\Delta u(t-d), \Delta u(t-d+1), \cdots, \Delta u(t-d+N_m-1)]^T$$

$$\Delta U(t-d-1) = [\Delta u(t-d-n_b), \Delta u(t-d-n_b+1), \cdots, \Delta u(t-d-1)]^T$$

$$F = [F_1(z^{-1}), F_2(z^{-1}), \cdots, F_N(z^{-1})]^T$$

$$Q = \begin{bmatrix} g_{n_b} & \cdots & g_1 \\ \vdots & \cdots & \vdots \\ g_{n_b}+N-1 & \cdots & gN \end{bmatrix}$$

Let $G_i(z^{-1}) = B(z^{-1})E_i(z^{-1}) = g_0 + g_1 z^{-1} + \cdots + g_{nb+i-1} z^{-(nb+i-1)}$, we can construct matrix P and Q by solving the Diophantine equation.

It should be noted that the item $\Delta U(t-d) = [\Delta u(t-d), \Delta u(t-d+1), \cdots, \Delta u(t-d+N_m-1)]^T$ in Equation (3) can not be obtained at the current time t. We assume $\Delta u(t-d+i) = 0, i = 0, 1, \cdots, N_m - 1$. This assumption is reasonable. On the one hand, the dynamic response of the future time can be reflected by using the past control actions and the previous process outputs. On the other hand, the item $y(t-1)$ is corresponding to the correction of the predicted model. That is $y(t-1) = y_m(t-1) + e(t-1)$, where $y_m(t-1)$ is the model output and $e(t-1)$ is the error between the process output and the model output.

2.3. Fuzzy Controller

The successful implementation of networked predictive fuzzy control relies on an assurance that the fuzzy controller must calculate all the pre-defined control actions in every sample time. So the fuzzy control algorithm should be simple and suitable to the real-time control. A real-time simplified fuzzy controller [16] is used to generate the future control actions. Unlike the ordinary fuzzy control, which uses error and error change at the present time to produce the current control action, the networked predictive fuzzy controller uses 'future error' and 'future error change' to derive 'future control actions.' The ordinary fuzzy control can be regarded as a special condition of the networked fuzzy control when the predictive step is equal to zero. This networked predictive fuzzy controller has two-input one-output. One input is the error e between the desired future output and the predicted output. The other one is the change of the error ec. The output of the fuzzy controller is the change of the future control action Δu. The membership functions of e and ec are adopted as triangular forms and the membership function of the output Δu is adopted discrete form as Figure 2 shown.

The 'Simplification' of the fuzzy controller relays on the inference process. For the typical two-input-one-output fuzzy controller, only four control rules are excited at each cycle time with the form 'If e is L_e and ec is L_{ec}, then Δu is $L_{\Delta u}$,' where L is the linguistic variables PB, PM, PS, ZE, NS, NM, NB. Due to the characteristic of the triangular membership function, e is at most belong to two membership functions μ_e^i and μ_e^{i+1}, ec is at most belong to two membership functions μ_{ec}^j ec and μ_{ec}^{j+1}, thus Δu has 2 × 2 combinations, that is four control rules:

if e is $L_e^{(i)}$ and if ec is $L_{ec}^{(j)}$ then Δu is $L_{\Delta u}^{(i,j)}$

if e is $L_e^{(i)}$ and if ec is $L_{ec}^{(j+1)}$ then Δu is $L_{\Delta u}^{(i,j+1)}$

if e is $L_e^{(i+1)}$ and if ec is $L_{ec}^{(j)}$ then Δu is $L_{\Delta u}^{(i+1,j)}$

if e is $L_e^{(i+1)}$ and if ec is $L_{ec}^{(j+1)}$ then Δu is $L_{\Delta u}^{(i+1,j+1)}$.

From Figure 2(c) shown, output Δu is adopted discrete form membership function. It is assumed that output domain has been divided into $c_m(k)$, m = 1, 2, ▫▫▫, n. Do minimum and maximum operator, calculate

$$\tilde{\mu}_{L_{\Delta u}^{(i,j)}}^{(m)} = \Delta(\mu_{L_e^{(i)}}, \mu_{L_{ec}^{(j)}}, \mu_{L_{\Delta u}^{(i,j)}}^{(m)})$$

$$\tilde{\mu}_{L_{\Delta u}^{(i,j+1)}}^{(m)} = \Delta(\mu_{L_e^{(i)}}, \mu_{L_{ec}^{(j+1)}}, \mu_{L_{\Delta u}^{(i,j+1)}}^{(m)})$$

$$\tilde{\mu}_{L_{\Delta u}^{(i+1,j)}}^{(m)} = \Delta(\mu_{L_e^{(i+1)}}, \mu_{L_{ec}^{(j)}}, \mu_{L_{\Delta u}^{(i+1,j)}}^{(m)})$$

$$\tilde{\mu}_{L_{\Delta u}^{(i+1,j+1)}}^{(m)} = \Delta(\mu_{L_e^{(i+1)}}, \mu_{L_{ec}^{(j+1)}}, \mu_{L_{\Delta u}^{(i+1,j+1)}}^{(m)})$$

(4)

and

$$\tilde{\tilde{\mu}}_{\Delta u}^{(m)} = \vee(\tilde{\mu}_{L_{\Delta u}^{(i,j)}}^{(m)}, \tilde{\mu}_{L_{\Delta u}^{(i,j+1)}}^{(m)}, \tilde{\mu}_{L_{\Delta u}^{(i+1,j)}}^{(m)}, \tilde{\mu}_{L_{\Delta u}^{(i+1,j+1)}}^{(m)})$$

(5)

where m = 1, 2, $\cdots$, n.

It is not easy to directly get the inference rules of the future; however, the typical dynamic of the second order system can be obtained ahead of time. Figure 3 presents the phase plane of the typical second order linear system with the x-axis standing for variable e and y-axis standing for variable ec. From this figure, it can be seen that the points in x-axis and y-axis are crucial. If these points are controlled very well, the control performance is guaranteed. So the main inference rules are deduced as shown in Table 1.

Thus, the incremental control actions

$$\Delta u(t+\eta) = \frac{\sum_{m=1}^{n} \tilde{\tilde{\mu}}_{\Delta u}^{(m)} * c_m}{\sum_{m=1}^{n} \tilde{\tilde{\mu}}_{\Delta u}^{(m)}}$$

(6)

and the predicted control actions

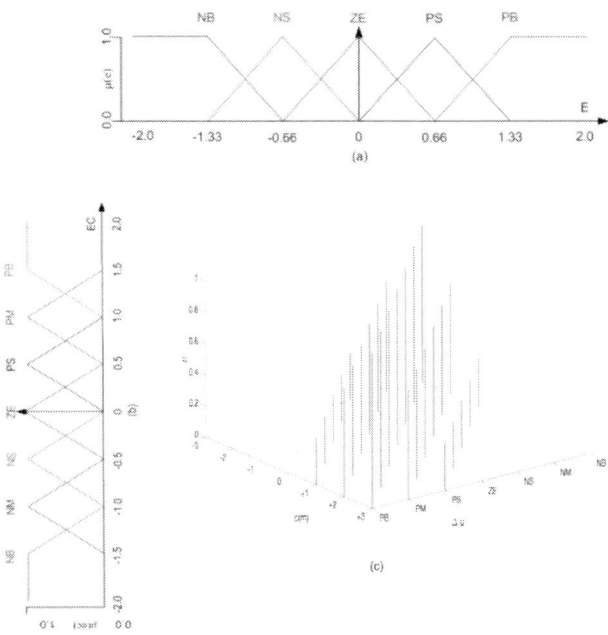

Figure 2. Membership functions of e, ec and Δu. (a) is the membership function of input e, (b) is the membership function of input ec, (c) is the discrete form membership function of output Δu.

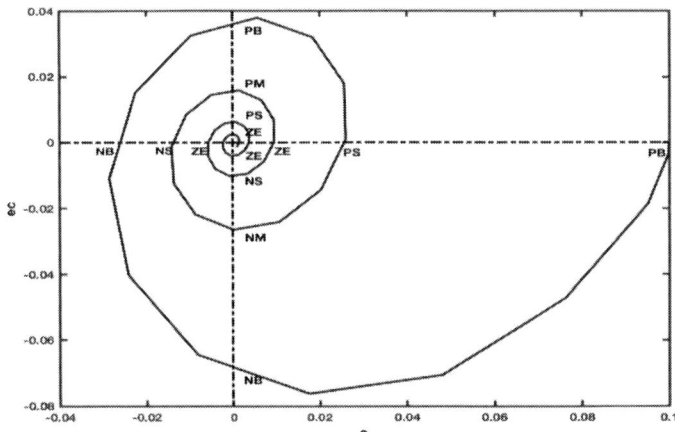

Figure 3. Phase plane and fuzzy control rules analysis of typical second order linear system.

$$u(t+\eta) = u(t+\eta-1) + \Delta u(t+\eta), \eta \text{ is integer and } 0 \leq \eta \leq N_u - 1 \quad (7)$$

can be given, where N_u is the control horizon. If e is the error and ec is the error change at present time, then Δu is the incremental control action at present time. If e and ec are the values of the future time, then the future incremental control actions can be derived. As paper [17–19] presents, variable domain implemented in the fuzzy control can greatly improve the control accuracy, which has successfully been applied to the control of a quadruple inverted pendulum [20] and the wing rock phenomenon [21]. This is achieved by the domain contraction and expansion. Domain contraction is equivalent to the increase of the control rules. Although the real-time algorithm has considered only 11 rules, together with the variable domain strategy, the fuzzy controller can acquire satisfactory control performance. The varies of domain can be achieved by multiplying flex factors $\alpha(e)$, $\beta(ec)$ and $\gamma(e, ec)$ of domains E, EC and ΔU, respectively. There are many different forms of flex factors. In this chapter, we adopt the exponential form of functions:

$$\alpha(e) = \left[\frac{|e|}{E}\right]^{\tau_1}, 0 < \tau_1 < 1$$

$$\beta(ec) = \left[\frac{|ec|}{EC}\right]^{\tau_2}, 0 < \tau_2 < 1$$

$$\gamma(e,ec) = \left[\left[\frac{|e|}{E}\right]^{\tau_1}\left[\frac{|ec|}{EC}\right]^{\tau_2}\right]^{\tau_3}, 0 < \tau_1, \tau_2, \tau_3 < 1$$

$$(8)$$

To summarized, the networked predictive fuzzy controller has eleven parameters to be designed. Four parameters are related to the model predictor. They are the order of the predictive model n_a and n_b, the predictive horizon N, and the model control horizon N_m. Seven parameters are belonging to the networked fuzzy controller. They are the control horizon N_u, the scaling gains K_e, K_{ec}, $K_{\Delta u}$ of error e, error change ec and incremental control action Δu, and the variable domain parameters τ_1, τ_2 and τ_3. The program steps for the networked predictive fuzzy control plus variable domain strategy are summarized below:

Step 1: Use Equation (3), calculate the future outputs $y_p(t|t)$, $y_p(t + 1|t)$, $\cdots$, $y_p(t + N - 1|t)$ of the controlled system according to the delayed output of the feedback channel and the previous control actions.

Step 2: Calculate the differences between the desired future outputs $r(t|t)$, $r(t + 1|t)$, $\cdots$, $r(t + N -1|t)$ and the model predicted values $y_p(t|t)$, $y_p(t+ 1|t)$, $\cdots$, $y_p(t+ N -1|t)$ to get $e(t|t)$, $e(t+ 1|t)$, $\cdots$, $e(t + N - 1|t)$ and $ec(t|t)$, $ec(t + 1|t)$, $\cdots$, $ec(t + N - 1|t)$.

Step 3: Adjust input and output domain using Equation (8) in terms of $e(t + \eta)$ and $ec(t + \eta)$.

Step 4: Calculate membership functions of input $e(t + \eta)$ and $ec(t + \eta)$ and output $\Delta u(t + \eta)$.

Step 5: Use minimum-maximum inference method [see Equations (4) and (5)].

Step 6: Calculate the predicted control actions $u(t + \eta)$ using Equations (6) and (7).

Step 7: Let $\eta = 0$ to $N_u - 1$, repeat step 3-6.

Step 8: Send the control actions $\Delta u(t), \Delta u(t + 1), \cdots, \Delta u(t + N_u - 1)$ with a packet to the plant side.

Step 9: Select the control action $u(t|t - k)$ and add to the controlled process.

Step 10: In the next sample time, repeat step 1-9.

Table 1. Main control rules of the fuzzy controller in NPFC

No.	Control rules
1	if e = PB and ec = ZE then Δu = PB
2	if e = ZE and ec = NB then Δu = NB
3	if e = NB and ec = ZE then Δu = NB
4	if e = ZE and ec = PB then Δu = PB
5	if e = PS and ec = ZE then Δu = PS
6	if e = ZE and ec = NM then Δu = NM
7	if e = NS and ec = ZE then Δu = NS
8	if e = ZE and ec = PM then Δu = PM
9	if e = ZE and ec = NS then Δu = NS
10	if e = ZE and ec = PS then Δu = PS
11	if e = ZE and ec = ZE then Δu = ZE

2.4. Network Delays Compensation

It is assumed that the network communication delay in the forward channel is not greater than the length of the predicted control horizon. To make use of the 'packet transmission' characteristic of the network, a string of future control actions which contain $u(t)$, $u(t+1)$, $\cdots$ $u(t+N_u-1)$ at sample time t are sent to the plant side at the same time. Then the control value from the latest control sequence available on the plant side is chosen as a control input of the plant to compensate for the forward channel delay. For example, if the latest control sequence on the plant side is

$$\begin{bmatrix} u(t-k \mid t-k) \\ u(t-k+1 \mid t-k) \\ \vdots \\ u(t-k+N_u-1 \mid t-k) \end{bmatrix} \qquad (9)$$

Then the output selected control signal will be

$$u(t) = u(t \mid t-k) \qquad (10)$$

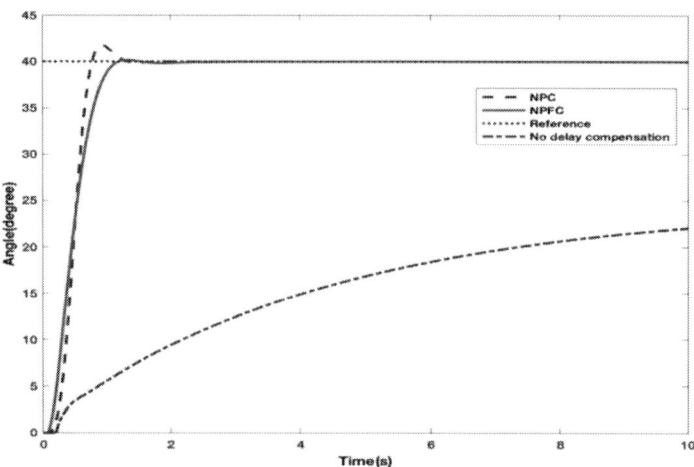

Figure 4. The step responses of NPC and NPFC with 1-step forward channel delay.

3. SIMULATIONS

Consider the servo control system as reference [2] shown. The system model with discrete form is as follows:

$$G(z^{-1}) = \frac{-0.00886z^{-1} + 1.268227z^{-2}}{1 - 1.66168z^{-1} + 0.6631z^{-2}} \tag{11}$$

where sample time is 0.04 second. Suppose there is one step delay in the forward channel, following Section 2.2 and Section 2.3, we design a linear model predictor and a real-time fuzzy controller. The parameters of the model predictor are: $n_a = 2$, $n_b = 2$, $N = 12$, $N_m = 10$. The parameters of the networked fuzzy controller are: $N_u = 10$, $K_e = 0.00125$, $K_{ec} = 0.02$, $K_{\Delta u} = 0.6$, $r_1 = 0.1$, $r_2 = 0.1$ and $r_3 = 0.01$. The NPC parameters are set to $N = 12$, $N_u = 10$, $\rho = 1500$. Figure 4 shows the control performance of NPFC and NPC. The dot line is the set-point. The solid line stands for the NPFC method. The dash line stands for NPC method. The dash-dot line stands for the NFPC method without delay compensation. From the figure, it can be seen that the NPFC can be regulated better than the NPC in control performance with rapid dynamic and small overshot. The delay compensation mechanism is very effective.

Suppose the case that six step delays exist in the forward channel. The NPFC controller parameters are adjusted as: $K_e = 0.0004$, $K_{ec} = 0.008$, $K_{\Delta u} = 0.08$, and the NPC parameters are set to $N = 25$, $N_u = 10$, $\rho = 100000$. To testify the control performance of the networked predictive fuzzy control method, the results of the NPC and the NPFC are presented in Figure 5. Through model prediction, fuzzy controller design and delay compensation, the NPFC presents very obviously better performance than NPC method. The rising time of NPFC is about 1.1 seconds while 1.5 seconds for NPC method. Moreover, NPC has 3.75% overshot while NPFC has nearly no overshot. When NPFC method not considers delay compensation, static errors can be seen in Figure 5. On the contrary, the dynamic response reaches steady state after 1.4 seconds when the delay compensator is acting.

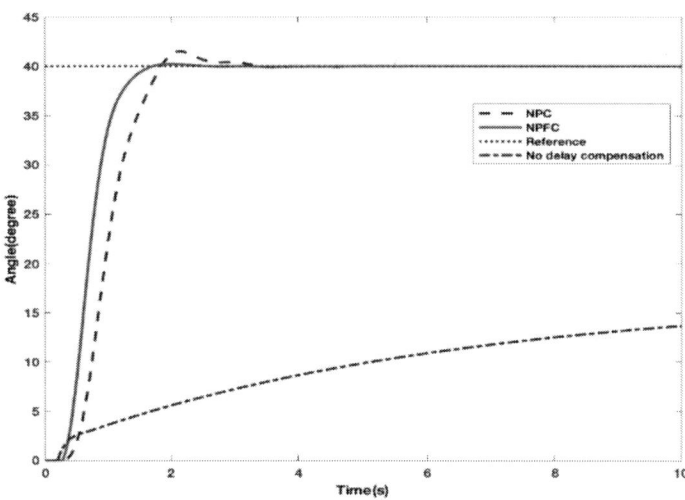

Figure 5. The step responses of NPC and NPFC with 6-step forward channel delay.

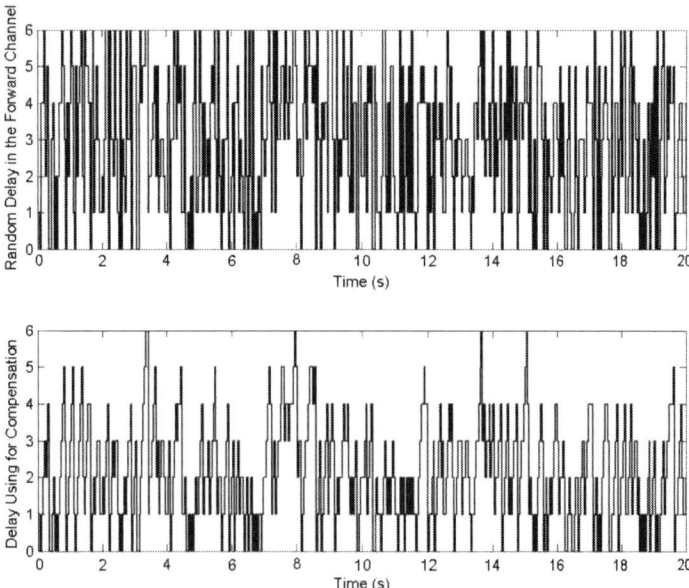

Figure 6. Random time delays in the NPFC.

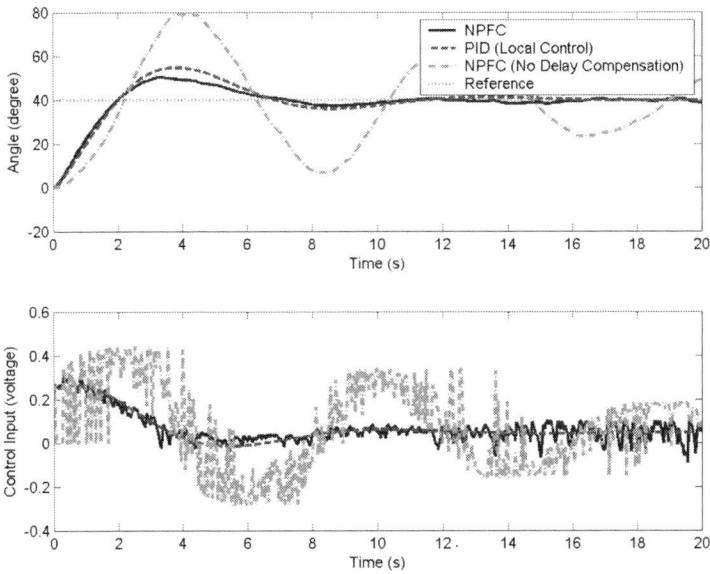

Figure 7. Control performance comparison with 0-step to 6-step random delays.

Let us consider the random delay conditions. Suppose there exists zero to six step random time delay in the forward communication channel (as shown in Figure 6) and one step delay in the backwards channel. To obtain the error and the change of error in the future time, a linear model predictor is designed. Its parameters are set: $n_a = 2$, $n_b = 2$, $N = 12$, $N_m = 10$. The model predictor generates predicted output of the plant model by using the delayed output of the actual plant and the historical control actions. Reference minus predicted output is equal to the input error of the fuzzy controller. Differential of the error is amount to the variable change of error. The error and change of error are selected as two inputs of the fuzzy controller. The parameters of the networked fuzzy controller are: $N_u = 10$, $K_e = 0.001$, $K_{ec} = 0.022$, $K_{\Delta u} = 0.18$, $\tau_1 = 0.01$, $\tau_2 = 0.001$ and $\tau_3 = 0.01$. Figure 7 shows the control performance of the NPFC algorithm together with the control input of each method. The dot line is the reference. The solid line stands for the NPFC method. The dash line represents the local PID controller. The dash-dot line stands for the NFPC method without delay compensation. From the figure, it can be seen that performance of the

NPFC is similar to the local PID controller at the same rising time. In each cycle, the predicted control actions are packed and sent from the controller side to the plant side. They are stored in the buffer as a series of candidate control actions. By comparing the time delay of each candidate control sequence, the latest one is selected. That is to say it has the minimum time delay. The original random time delay is shown in the upper diagram of the Figure 6, and the time delay adopted in the compensation system is shown in the lower diagram. Without the delay compensation, the curve is depicted with the dash-dot line (yellow one). From the solid line (blue one) and the dash-dot line (yellow one), it can be seen that the control performance has been improved. This fact has proved that the random time delay compensation mechanism is very effective.

CONCLUSION

This chapter proposes a network predictive fuzzy controller structure. By solving the Diophantine equation, the linear model predictor generates a series of predictive outputs of the controlled system according to the outputs and the control sequence in the past time. The errors between the desired future outputs and the predictive outputs from the linear model predictor and the error change are used to design a real-time fuzzy controller. So a series of future control sequence is produced in the controller side. By selecting the appropriate control sequence in the plant side, the fixed and random delays in the forward channel are compensated. Because NPFC has more parameters which can be regulated, the control performance can be adjusted better than NPC and PID method.

ACKNOWLEDGMENTS

This work is supported by the Science and Technology Program of Beijing Municipal Education Commission (KM201811417001, KM202011417004), the Beijing Natural Science Foundation-Beijing Municipal Education Commission Joint Fund (KZ201811417048), the

Beijing Natural Science Foundation-Rail Transit Joint Fund (L191006) and the open research fund of the State Key Laboratory for Management and Control of Complex Systems (20210111).

REFERENCES

[1] Liu G. P., Rees D., Chai S. C., Nie X. Y. (2005). Design, simulation and implementation of netowrked predictive control systems, *Measurement and Control*, 38,17-21.
[2] Liu G. P., Mu J. X., Rees D., Chai S. C. (2006). Design and stability analysis of networked control systems with random communication time delay using the modified MPC, *International Journal of Control*, 79(4),288-297.
[3] Mahmoud M. S., Saif A. A. (2012). Robust quantized approach to fuzzy networked control systems, *IEEE Journal on Emerging and Selected Topics in Circuits and Systems*, 2(1), 71-81.
[4] Du F., Qian Q. Q. (2008). The research of heterogeneous networked control systems based on modify smith predictor and fuzzy adaptive control, *IEEE International Conference on Fuzzy Systems*.
[5] Tang P. L., De S. C. W. (2006). Compensation for transmission delays in an ethernet-based control network using variable-horizon predictive control, *IEEE Transactions on Control Systems Technology,* 14(4), 707-718.
[6] Jia X. C., Zhang D. W., Zheng L. H., Zheng N. N. (2008). Modeling and stabilization for a class of nonlinear networked control systems: A T-S fuzzy approach, *Progress in Natural Science*, 18(8), 1031-1037.
[7] Jiang X. F., Han Q. L. (2008). On designing fuzzy controllers for a class of nonlinear networked control systems, *IEEE Transactions on Fuzzy System*, 16(4), 1050- 1060.
[8] Hajebi P., Almodarresi S. M. T. (2012). Online adaptive fuzzy logic controller using neural network for Networked Control Systems, *International Conference on Advanced Communication Technology.*
[9] Ren C. Q., Wu P. D., Wang X. F., Ma S. Y., Chen Z. L. (2002). A study on the forecast arithmatic of hydraulic telecontrol system

based on internet, *Journal of Beijing Institute of Technology*, 22(1), 85-89.
[10] Zhen W., Xie J. Y. (2002). On-line delay-evaluation control for networked control systems, *IEEE Conference on Decision and Control.*
[11] Zhang Y. Y., Zhang J. L., Luo X. Y., Guan X. P. (2013). Faults detection for networked control systems via predictive control, *International Journal of Automation and Computing*, 10(3),173-180.
[12] Tang X. M., Ding B. C. (2012). Design of networked control systems with bounded arbitrary time delays, *International Journal of Automation and Computing*, 9(2),182- 190.
[13] Li P. F., Yan X. P., Qiu L. P., Zhou Q. Y. (2009). Study on predictive fuzzy control of great inertia system based on grey model, *2009 Second International Conference on Intelligent Computation Technology and Automation.*
[14] Hu J. Q., Rose E. (1997). Predictive fuzzy control applied to the sinter strand process, *Control Engineering Practice*, 5(2), 247-252.
[15] Tong S. W., Liu G. P. (2007). Design and Simulation of Fuel Cell Networked Predictive Fuzzy Control Systems, *Proceedings of the 26th Chinese Control Conference.*
[16] Tong S. W., Liu G. P. (2008). Real-time simplified variable domain fuzzy control of pem fuel cell flow systems, *European Journal of Control*, 14(3), 223-233.
[17] Li H. X. (1995). To see the success of fuzzy logic from mathematical essence of fuzzy control, *Fuzzy Systems and Mathematics*, 9(4), 1-14.
[18] Oh S. Y., Park D. J. (1995). Self-tuning controller with variable universe of discourse, IEEE International Conference on Systems, *Man and Cybernetics.*
[19] Li H. X. (1999). Variable universe adpative fuzzy controller, *Science in China (Series E)*, 20(1), 32-42.
[20] Li H. X., Miao Z. H., Wang J. Y. (2002). Variable universe adaptive fuzzy control on the quadruple inverted pendulum, *Science in China (Series E)*, 45(2), 213-224.
[21] Liu Z. L., Su C. Y., Svoboda J. (2004). Control of wing rock phenomenon with a variable universe fuzzy controller, *Proceeding of the American Control Conference.*

In: Networked Control Systems
Editors: S. Tong and D. Qian
ISBN: 978-1-53619-892-8
© 2021 Nova Science Publishers, Inc.

Chapter 2

DESIGN OF A DATA-BASED NETWORKED TRACKING CONTROL SYSTEM

Shiwen Tong[1,3,*] *and Dianwei Qian*[2]

[1]College of Robotics, Beijing Union University, Beijing, China
[2]School of Control and Computer Engineering,
North China Electric Power University, Beijing, China
[3]State Key Laboratory for Management and Control of Complex Systems, Institute of Automation, Chinese Academy of Sciences, Beijing, China

ABSTRACT

Tracking control is a very challenging problem in the Networked Control System (NCS), especially for the process with blurred mechanism and where only input-output data are available. This chapter has proposed a data-based design approach for the Networked Tracking Control System (NTCS). The method utilizes the input-output data of the controlled process to establish a predictive model with the help of Fuzzy Cluster Modelling (FCM) technology. Then, the deduced error and error change in the future are treated as inputs of a Fuzzy Sliding Mode Controller (FSMC) to obtain a string of future control actions. These candidate control

---

* Corresponding Author's E-mail: shiwen.tong@buu.edu.cn.

actions in the controller side are delivered to the plant side. Thus, the network induced time delays are compensated by selecting appropriate control action. Simulation outputs prove the validity of the proposed method.

Keywords: networked tracking control, fuzzy cluster modeling, fuzzy sliding mode control, delay compensation

1. INTRODUCTION

Control theories have been enriched with the development of the network where sensor, controller and actuator are connected to form a close-loop structure, called Networked Control System (NCS) [1]. Networked Tracking Control (NTC), as a branch of the networked control theories, has gradually received attention in recent years. Until now, there are two streams of studies for the NTC. One is focusing on the model predictive control [2] [3]. The other is concentrating on the robust control [4] [5] [6] [7]. The representative of the former is the research of Pang et al. [3]. They treated the tracking error as an additional state and considered the delay, packet disorder and packet dropout by using the round-trip time-delay and augment method. The typical research of the latter is from the scholars Gao et al. [4], Wang et al. [5] and Qiu et al. [6]. They discussed the NTC problem in the H∞ sense and considered the different phenomenons such as time delay, packet dropout, constant and varying sampling conditions.

Different from the set-point control of the NCS, NTC is more challenging since the accuracy is not the only consideration, but the rapidity is also deserved to pay attention. NTC system requires the process output of the controlled system to track the variations of the references, in timely and accurate manner [8]. Furthermore, it can be seen from the recent literatures that most researches are model-based design [2] [3] [4] [5]. The structure and the parameters of the process model in the network must be known before the controller design. However, the mechanism of some process may not be clear in a practical situation, that is to say, they are systems with blurred mechanism. This motivates us to develop a data-based design approach for the NTC system.

Design of a Data-Based Networked Tracking Control System 21

The rest of the chapter is organized as follows: Section 2 presents the core idea and the structure of the Data-based Networked Tracking Control (DNTC) system. Section 3 gives the detailed design procedure. Simulations and comparisons are presented in Section 4 and then conclusion is drawn.

2. CORE IDEA AND STRUCTURE OF THE DNTC SYSTEM

As we all know, due to the media sharing characteristic of network, time delay, data disorder and data dropout, etc. will degrade the control performance of the NCS, these factors can even make the process unstable. Among these influential factors, time delay is the fundamental problem. Some other factors such as data disorder and data dropout can be regarded as a special case of the time delay issue. Therefore, the DNTC approach must consider the compensation of the time delay and make the control algorithm accurate and timely. In summary, the DNTC system should solve the following issues:

- The contradiction between the timeliness of the tracking control and the inevitable time delay phenomenon in the network.
- Obtaining the future control actions at current time.
- Active compensation of the network induced time delay.
- Design NTC system without knowing the model, instead, only input-output data of the process are available.

To solve the first problem, 'prediction idea' is adopted. If a string of future control actions can be predicted and transferred from the controller side to the plant side, the time delay issue is solved by using a certain future control action instead of the delayed one [9]. To solve the second problem, the idea of 'coping with shifting events by sticking to a fundamental principle' is adopted. If the relatively unchanged relationship between the future errors/error changes and the future control actions is established, the second problem is readily solved. This can be realized by abstracting a fuzzy sliding mode controller which has broad applications as a variable structure controller [10] [11]. To solve the third problem, 'packet transfer' characteristic and time stamps of the network are implemented. The time delay can be actively compensated on the plant side by selecting appropriate control action from the data packet, transferred from the controller side, according to the measured delay information calculated by subtracting the time stamp recorded in the data packet from the present time [12]. To solve the last problem, Fuzzy Cluster Modelling (FCM) technology is used. This technology can approximate a linear or nonlinear system with some local sub-models, only according to the input-output data of the process, by overlapping these individual sub-models [13].

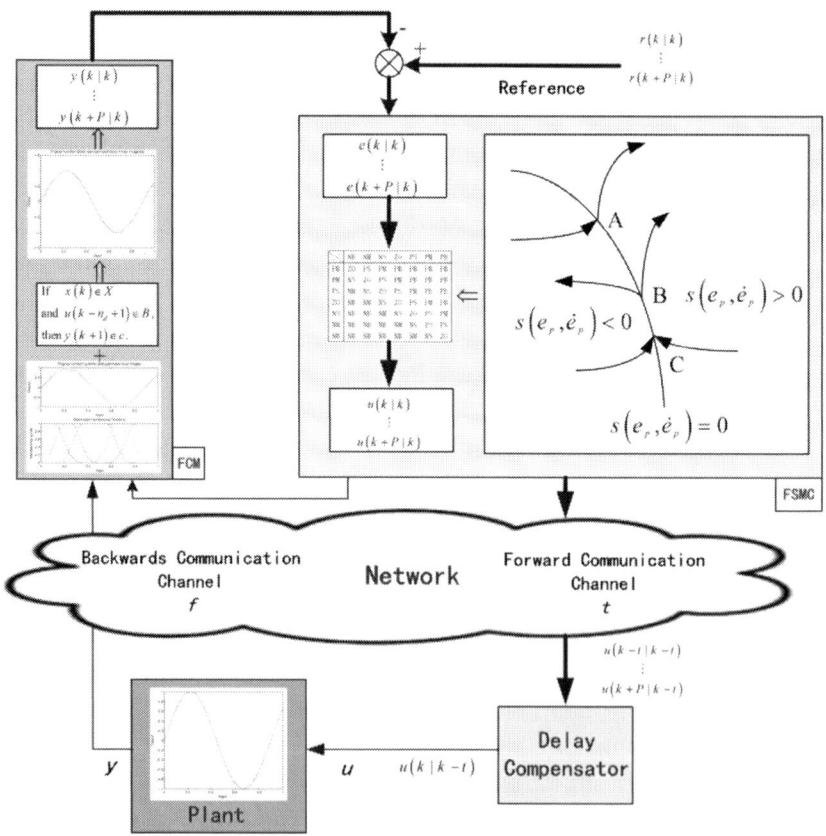

Figure 1. Structure of DNTC system.

The structure of the DNTC system is presented in Figure 1 which is composed of three key components: the predictor, the Fuzzy Sliding

Mode Controller (FSMC) and the delay compensator. The system can be divided into two parts: the controller side and the plant side. They are connected in a networked form including the forward channel and the backward channel. On the controller side, the predictor is used to generate the future outputs from the input-output data of the process, with fuzzy cluster modelling technology. The error and the change of error between the desired references and the predicted outputs of the process are fed to the FSMC to produce the future control actions. Then, the future control actions are packed, together with the time stamps, and sent to the plant side. On the plant side, the delay can be compensated by selecting the appropriate control action. To solve the problems step by step, only the forward communication channel delay is considered in this chapter.

3. THE DETAILED DESIGN

3.1. Predictive Model Generated from Input-Output Data

Suppose the controlled process is to be described by a T-S fuzzy model.

R_i: If $y(k)$ is A_i, $y(k-1)$ is G_i and $u(k)$ is B_i

then

$$y(k+1) = a_i y(k) + g_i y(k-1) + b_i u(k), \, i = 1,2,\cdots,K \quad (1)$$

where fuzzy sets A_i, G_i and B_i can be described by the membership functions from GK algorithm, and a_i, g_i, b_i can be estimated by the least square identification using the input-output data of the controlled process.

The GK fuzzy clustering algorithm [14] is an optimization of the cost function with the form

$$\min_{(U,V,A)} \left\{ J(Z;U,V,A) = \sum_{i=1}^{c}\sum_{k=1}^{N}(\mu_{ik})^m D_{ikA_i}^2 \right\} \quad (2)$$

$$D_{ikA_i}^2 = (z_k - v_i)^T A_i (z_k - v_i)$$

where $Z = \{z_k | k = 1, 2, \cdots, N\}$ is an observation vector which consists of the input-output data pairs of the controlled process. $A = [A_1, \cdots, A_c]$ is a set of c norm-inducing matrices, m is a fuzziness parameter of the clusters, $U = \mu_{ik}$ is a $c \times N$ matrix representing the fuzzy partition, $V = [v_1, \ldots, v_c]$ is a vector of cluster prototypes (centers), D^2_{ikAi} is a squared inner-product distance norm. GK algorithm is an extension of the standard fuzzy c-means algorithm by employing an adaptive distance norm. It is realized by defining a norm-inducing matrix A_i for each cluster. Since unlimited A_i will affect the feasible solution of the optimization, a constraint is applied to the determinant of A_i.

$$|A_i| = \rho_i, \rho_i > 0, \forall i. \quad (3)$$

Using the Lagrange multiplier method, the following expression for A_i is obtained

$$A_i = [\rho_i \det(F_i)]^{1/n} F_i^{-1} \quad (4)$$

where F_i is the fuzzy covariance matrix of the ith cluster determined by

$$F_i = \frac{\sum_{k=1}^{N} \mu_{ik}^m (z_k - v_i)(z_k - v_i)^T}{\sum_{k=1}^{N} \mu_{ik}^m} \quad (5)$$

Then, the point-wise membership functions can be defined by

$$\mu_{ik} = \frac{1}{\sum_{j=1}^{c}(D_{ikA_i} / D_{jkA_i})^{2/(m-1)}} \quad (6)$$

In order to obtain a uniform predictive model, the point-wise membership functions must be translated into some mathematical functions. However, the point-wise membership functions are usually substituted by the piece-wise exponential membership functions. The characteristics of these alternative membership functions lie in that they have implicit antecedent variables y(k), y(k−1) or u(k) and cannot be directly used in the controller design. To solve this kind of problem, two-layer feed-forward networks with sigmoid hidden neurons and linear output neurons are utilized to transform the implicit form of membership functions into the explicit form of membership functions

$$\left.\begin{array}{l} \mu_{A_i} = f(y(k)) \\ \mu_{G_i} = f(y(k-1)) \\ \mu_{B_i} = f(u(k)) \end{array}\right\} \Rightarrow \left.\begin{array}{l} \mu_{A_i} = Net_{A_i} \cdot y(k) \\ \mu_{G_i} = Net_{G_i} \cdot y(k-1) \\ \mu_{B_i} = Net_{B_i} \cdot u(k) \end{array}\right\} \quad (7)$$

Then, the antecedent membership functions μ_{Ai}, μ_{Gi} and μ_{Bi} can be described by $\mu_{Ai} = Net_{Ai} \cdot y(k)$, $\mu_{Gi} = Net_{Gi} \cdot y(k-1)$ and $\mu_{Bi} = Net_{Bi} \cdot u(k)$. Thus, the implicit antecedent membership functions have been transformed into the explicit functions and the inference results of fuzzy T-S model can be easily obtained by

$$y(k+1) = \frac{\sum_{i=1}^{k} \mu_{A_i} \cdot \mu_{G_i} \cdot \mu_{B_i} (a_i y(k) + g(k-1) + b_i u(k))}{\sum_{i=1}^{k} \mu_{A_i} \cdot \mu_{G_i} \cdot \mu_{B_i}}$$

$$= \frac{\sum_{i=1}^{k} (Net_{A_i} \cdot y(k))(Net_{G_i} \cdot y(k-1))(Net_{B_i} \cdot u(k))(a_i y(k) + g_i y(k-1) + b_i u(k))}{\sum_{i=1}^{k} (Net_{A_i} \cdot y(k))(Net_{G_i} \cdot y(k-1))(Net_{B_i} \cdot u(k))} \quad (8)$$

$$= \frac{\sum_{i=1}^{k} Net_{A_i} \cdot Net_{G_i} \cdot Net_{B_i} (a_i y(k) + g_i y(k-1) + b_i u(k))}{\sum_{i=1}^{k} Net_{A_i} \cdot Net_{G_i} \cdot Net_{B_i}}$$

and the estimation of the consequent parameters a_i, g_i and b_i can be regarded as the least square estimation issue with the form

$$Y = X \cdot \theta + \xi \quad (9)$$

where $Y = [y(k), y(k-1), \cdots, y(k-N+1)]^T$ is the output matrix, $X = [y(k-1), y(k-2), u(k-1); y(k-2), y(k-3), u(k-2); \cdots; y(k-N), y(k-N-1), u(k-N)]$ is the measurement matrix, $\theta = [a_1, g_1, b_1; a_2, g_2, b_2; \cdots; a_c, g_c, b_c]^T$ is the parameter matrix and ξ is the noise matrix. The solution of the least square problem is

$$\theta = (X^T X)^{-1} X^T Y \tag{10}$$

Then, one step predictive equation is obtained by Equation (8)

$$y_p(k+1) = \hat{a}y(k) + \hat{g}y(k-1) + \hat{b}u(k) \tag{11}$$

where

$$\hat{a} = \left(\sum_{i=1}^{k} NetA_i \cdot NetG_i \cdot NetB_i \cdot a_i\right) / \left(\sum_{i=1}^{k} NetA_i \cdot NetG_i \cdot NetB_i\right),$$

$$\hat{g} = \left(\sum_{i=1}^{k} NetA_i \cdot NetG_i \cdot NetB_i \cdot g_i\right) / \left(\sum_{i=1}^{k} NetA_i \cdot NetG_i \cdot NetB_i\right),$$

$$\hat{b} = \left(\sum_{i=1}^{k} NetA_i \cdot NetG_i \cdot NetB_i \cdot b_i\right) / \left(\sum_{i=1}^{k} NetA_i \cdot NetG_i \cdot NetB_i\right).$$

$y_p(k+2)$, $y_p(k+3)$, ..., $y_p(k+P)$, where P is the predictive step, $P \geq 0$, can be obtained by recurrence.

$$\begin{aligned} y_p(k+2) &= \hat{a}y(k+1) + \hat{g}y(k) + \hat{b}u(k+1) \\ y_p(k+3) &= \hat{a}y(k+2) + \hat{g}y(k+1) + \hat{b}u(k+2) \\ &\vdots \\ y_p(k+P) &= \hat{a}y(k+P-1) + \hat{g}y(k+P-2) + \hat{b}u(k+P-1) \end{aligned} \tag{12}$$

To increase the prediction accuracy, predicted values at sample time $k + 1$, $k + 2$, ..., $k + P$ are corrected by the difference between actual process output $y(k)$ and the predicted output $y_p(k)$, i.e.,

$$y_p(k+1) = \hat{a}y(k) + \hat{g}y(k-1) + \hat{b}u(k) + \alpha_1(y(k) - y_P(k))$$
$$y_p(k+2) = \hat{a}y(k+1) + \hat{g}y(k) + \hat{b}u(k+1) + \alpha_2(y(k) - y_P(k))$$
$$y_p(k+3) = \hat{a}y(k+2) + \hat{g}y(k+1) + \hat{b}u(k+2) + \alpha_3(y(k) - y_P(k))$$
$$\vdots$$
$$y_p(k+P) = \hat{a}y(k+P-1) + \hat{g}y(k+P-2) + \hat{b}u(k+P-1) + \alpha_P(y(k) - y_P(k)) \quad (13)$$

where $\alpha_i \leq 1$ is corrective coefficient, the bigger the α is, the stronger the corrective action will be. To simplify the design and integrate with the local control, we let the Prediction Horizon (PH) be equal to the Control Horizon (CH), $PH = CH = P + 1$.

3.2. Design of FSMC

Let sample time be T and predictive step be P, the predicted error e_p and error change de_p can be written by

$$e_p(k+P) = r(k+P) - y_p(k+P)$$
$$de_p(k+P) = \frac{e_p(k+P) - e_p(k+P-1)}{T} \quad (14)$$

Switching functions are designed as

$$s_p(k+P) = \psi \cdot e_p(k+P) - de_p(k+P)$$
$$ds_p(k+P) = s_p(k+P) - s_p(k+P-1) \quad (15)$$

We adopt two-input one-output form of fuzzy controller. The inputs of the fuzzy controller are S_p and $\acute{S}_p$ which are fuzzy variables of $s_p(k+P)$ and $ds_p(k+P)$, respectively. The output of the fuzzy controller is ΔU_p which is the fuzzy variable of the incremental control action Δu. Define fuzzy sets: PB = Positive Big, PM=Positive Medium, PS = Positive Small, NS = Negative Small, NM=Negative Medium, NB = Negative Big and ZO = Zero, the variables S_p, $\acute{S}_p$ and ΔU_p are partitioned into seven fuzzy sets, respectively.

$$S_p = \{NB, NM, NS, ZO, PS, PM, PB\}$$
$$\acute{S}_p = \{NB, NM, NS, ZO, PS, PM, PB\}$$
$$\Delta U_p = \{NB, NM, NS, ZO, PS, PM, PB\} \tag{16}$$

The domains of S_p, $\acute{S}_p$ and ΔU_p are all denoted by {−3, −2, −1, 0, +1, +2, +3} and the normal membership functions are selected for them (See Figure 2). K_{Sp}, $K_{\acute{S}p}$ and $K_{\Delta Up}$ are scaling gains of $\acute{S}_p$, $\acute{S}_p$ and ΔU_p, respectively. Adopting the typical form of control rules 'If S_p is A and $\acute{S}_p$ is B, then ΔU_p is C', the control actions can be obtained by satisfying the condition $S_p \cdot \acute{S}_p < 0$. Thus, the fuzzy control rules are determined by Table 1.

The predicted control actions $\Delta u_p(k + P)$ are determined using the defuzzification method of the center of gravity. Thus, the control actions can be calculated by

$$u_p(k+P) = u_p(k+P-1) + \Delta u_p(k+P) \tag{17}$$

3.3. Delay Compensation

Suppose the delay in the communication channel is not greater than the control horizon. Due to the 'packet transmission' characteristic of the internet, the string of future control actions $u(k)$, $u(k + 1)$,□□□ $u(k + P)$ at sample time k can be packed together and sent to the plant side from the controller side.

Table 1. Fuzzy Control Rules

$S_p \backslash \acute{S}_p$	NB	NM	NS	ZO	PS	PM	PB
PB	ZO	PS	PM	PB	PB	PB	PB
PM	NS	ZO	PS	PM	PB	PB	PB
PS	NM	NS	ZO	PS	PM	PB	PB
ZO	NB	NM	NS	ZO	PS	PM	PB
NS	NB	NB	NM	NS	ZO	PS	PM
NM	NB	NB	NB	NM	NS	ZO	PS
NB	NB	NB	NB	NB	NM	NS	ZO

Then, the candidate control action can be selected from the newest control sequence. Thus, the time delay in the communication channel is compensated [15]. For example, if the newest control sequence on the plant side is

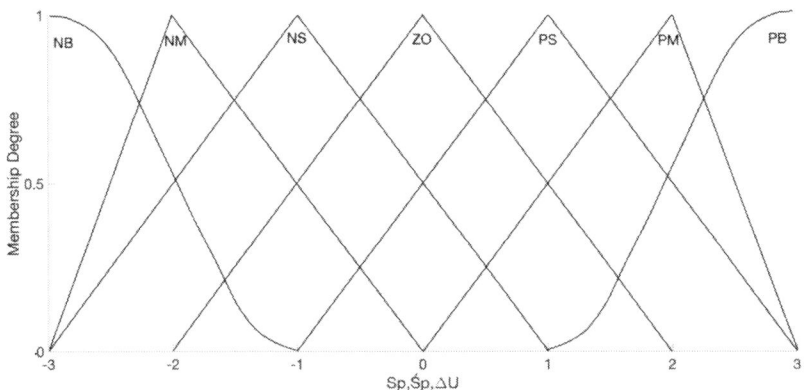

Figure 2. Membership functions definition.

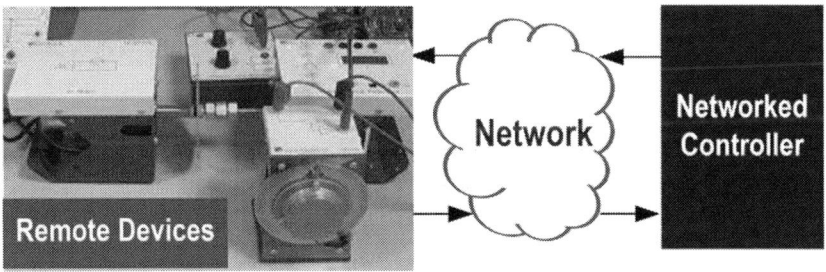

Figure 3. Networked control of the servo control system (SCS).

$$\begin{bmatrix} u(k-t \mid k-t) \\ u(k-t+1 \mid k-t) \\ \vdots \\ u(k-t+P \mid k-t) \end{bmatrix}$$

(18)

Then, the selected control signal will be

$$u(k) = u(k \mid k-t)$$

(19)

4. SIMULATIONS

As we all know, tracking control is often utilized in the motion control such as piezoelectric positioning stage [16], robots [17] and high-accuracy motion-systems driven by Doublesided Linear Switched Reluctance Machine (DLSRM) [6], linear motor [18] [19] or Direct Current (DC) motors [20] [21], etc. Among them, most of the controls are model-based (see [17], [18] and [20]). In this chapter, a data-based design has been proposed. To test the design, two examples are studied. One is for a Servo Control System (SCS). The other is for a nonlinear system: One-link Manipulator (OM).

4.1. Example I

The typical servo control system, which consists of dc motor, load plate and angle sensor, is considered as a simulation platform to test the proposed control method (see Figure 3). The control system is designed drive the load plate to a pre-set angle. The system is modeled by [22], which describes the relationship between the control input (voltage) and the angle position (degree).

$$G(z^{-1}) = \frac{-0.0086z^{-1} + 1.268227z^{-2}}{1 - 1.66168z^{-1} + 0.6631z^{-2}} \quad (20)$$

where sample time is 0.04 second.

The multi-sinusoidal signals with 0.4 amplitude, 0.06π, 0.012π, 0.03π rad/sec frequencies are selected as excitation sources to generate the data set for fuzzy cluster modelling. They are divided into two halves. One half is for fuzzy cluster modeling, and one half is for validation. The number of clusters is defined by $c = 3$, the fuzziness parameter is denoted by $m = 2$, and the termination criterion is 0.01. Then the three individual local models are constructed by GK algorithm.

R1: If $y(k)$ is A_1, $y(k-1)$ is G_1 and $u(k)$ is B_1
then $y(k+1) = 1.5028y(k) - 0.5056y(k-1) + 1.8415u(k)$
R2: If $y(k)$ is A_2, $y(k-1)$ is G_2 and $u(k)$ is B_2 (21)
then $y(k+1) = 2.3083y(k) - 1.3062y(k-1) - 0.9378u(k)$
R3: If $y(k)$ is A_3, $y(k-1)$ is G_3 and $u(k)$ is B_3
then $y(k+1) = 1.4374y(k) - 0.4399y(k-1) + 1.8998u(k)$

Thus, the basic predictive model is obtained.

$$y_p(k+1) = 1.6987y(k) - 0.7000y(k-1) + 1.0844u(k) \qquad (22)$$

From Equation (22), the predicted values $y_p(k), y_p(k+2), y_p(k+3), \ldots, y_p(k+P)$ are calculated by recurrence method. To increase the prediction accuracy, a correcting item $a_i(y(k) - y_p(k))$, $i = 1, 2, \ldots, P$ is added to each predicted value. In the Example I, two cases are considered: one is the local control of DNTC, the other is compared with PID controller for the system with 3-step delay in the forward communication channel. The fuzzy sliding mode controller is designed as Section 3.2 presented.

Case I

When the predictive step P is equal to zero, Data-based Networked Tracking Control (DNTC) has become Local Data-based Tracking Control (LDTC). That means there is no delay in the forward communication channel. Figure 4 shows the control performance of this case. The solid line in red color stands for the reference-a sine signal, and the dashed line in blue color is the DNTC (No Delay) output. The controller parameters are set as $K_{Sp} = 0.1$, $K_{\dot{S}p} = 0.1$, $K_{\Delta Up} = 0.015$, $\psi = 10$, $PH = CH = 1$. It can be seen that the output of the DNTC can follow the variations of the reference signal

Case II

In this case, we compare the DNTC method with regular PID controller considering three steps of delays in the forward communication channel. Figure 5 shows the control performance.

The solid line in red color represents the reference sine signal, the dashed line in blue color is the output of DNTC, and the dash-dot line in green color is the PID controller. The parameters of the DNTC controller are set as $K_{Sp} = 0.1$, $K_{\hat{S}p} = 0.1$, $K_{\Delta Up} = 0.01$, $\psi = 10$, $PH = CH = 6$, $a_i = 0.9$, $i = 1, \square\square\square, 6$ and the parameters of PID controller are defined by proportional coefficient $K_p = 0.01$, integral coefficient $K_i = 0.001$, filter coefficient $N = 100$. It can be seen that the DNTC method displays rapid response and high accuracy compared with the regular PID controller

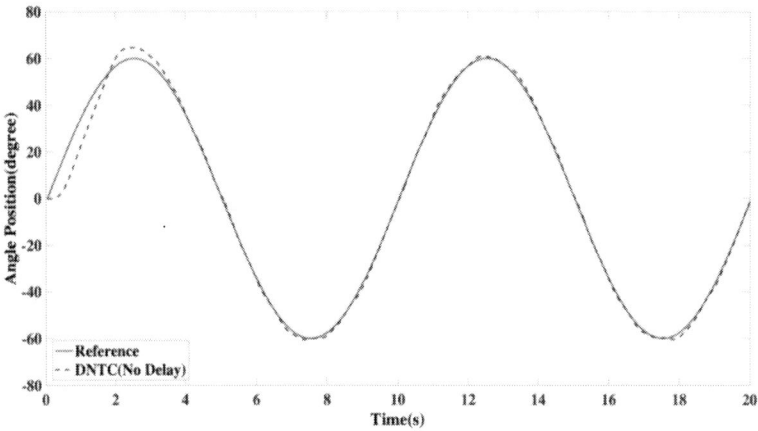

Figure 4. The control performance of DNTC (No Delay) for SCS.

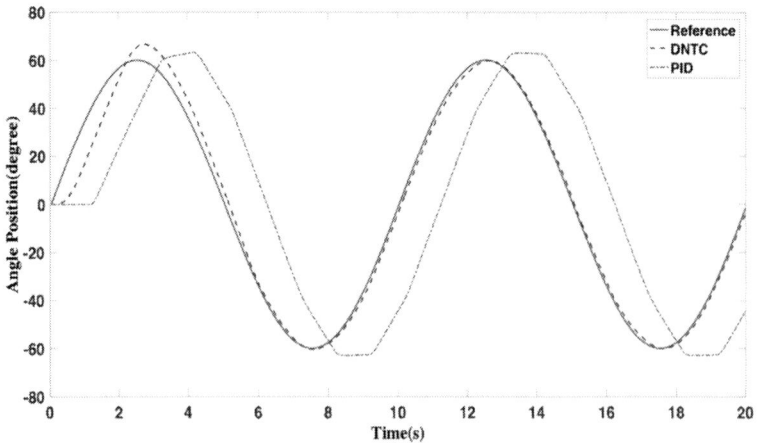

Figure 5. Control Performance of DNTC (3-step delay) for SCS Compared with PID.

4.2. Example II

To further demonstrate the validity of the proposed method, a nonlinear One-link Manipulator (OM) system is introduced (see Figure 6). It can be governed by the following nonlinear dynamics [23]:

$$\mathcal{D}\ddot{q} + \mathcal{B}\dot{q} + \mathcal{N}\sin(q) = \vartheta$$
$$\mathcal{M}\dot{\vartheta} + \mathcal{H}\vartheta = u - \mathcal{L}\dot{q} \tag{23}$$

where $\ddot{q}$, $\dot{q}$ and q are the link acceleration, velocity and position, respectively. ϑ stands for the torque generated by the electrical subsystem. u represents the input. The mechanical inertia $D = 1$ kg m², the coefficient of viscous friction at the joint $B = 1$ Nm s/rad, the armature resistance $H = 1$ Ω, the armature inductance $M = 0.1H$, and the back electromotive force coefficient $L = 0.2$ Nm/A.

Let sample time be 0.3 second. The output is the link position. The number of clusters is defined by $c = 3$, the fuzziness parameter is denoted by $m = 2$, and the termination criterion is 0.01. Then, the three individual local models are constructed by GK algorithm according to the input-output data generated from the controlled process after complete excitation.

R1 : If $y(k)$ is A_1, $y(k - 1)$ is G_1 and $u(k)$ is B_1
then $y(k+1) = 2.0407y(k) - 1.3808y(k-1) + 0.41535u(k)$
R2 : If $y(k)$ is A_2, $y(k - 1)$ is G_2 and $u(k)$ is B_2 (24)
then $y(k+1) = -0.0441y(k) + 0.577y(k-1) + 0.1935u(k)$
R3 : If $y(k)$ is A_3, $y(k - 1)$ is G_3 and $u(k)$ is B_3
then $y(k + 1) = 2.9682y(k) - 1.3424y(k-1) - 0.4147u(k)$

Therefore, the basic predictive model is obtained

$$y_p(k+1) = 1.6441y(k) - 0.7118y(k-1) + 0.0703u(k) \tag{25}$$

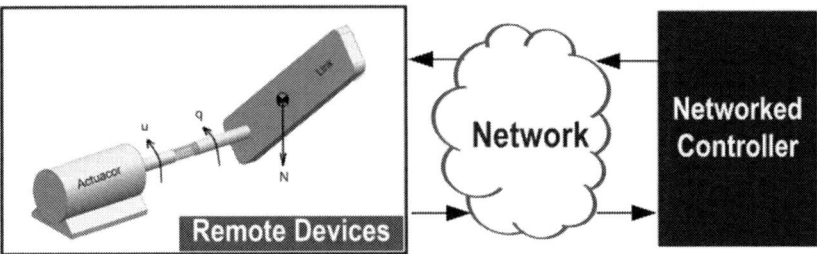

Figure 6. Networked control of the OM.

and the fuzzy controller can be designed following the instructions of Section 3

In this example, we compare the DNTC method with regular PID controller considering ten steps of delays in the forward communication channel. Figure 7 shows the control performance. The solid line in red color represents the reference square wave signal, the dashed line in blue color is the output of DNTC, and the dash-dot line in green color is the PID controller. The parameters of the DNTC controller are set as $K_{Sp} = 22$, $K_{\dot{S}p} = 2.8$, $K_{\Delta Up} = 0.9$, $\psi = 4.8$, $PH = CH = 12$, $\alpha_i = 0.9$, $i = 1, \square\square\square$ 10 and the parameters of PID controller are defined by proportional coefficient $K_p = 2.5$, integral coefficient $K_i = 4.5$, derivative coefficient $K_d = 2.8$, filter coefficient $N = 100$. It can be seen that ten steps of delays cannot be well handled by a PID controller. However, the DNTC method displays rapid response and high accuracy compared with the regular PID controller.

5. DISCUSSIONS

In fact, for many processes, the system (1) can also be simplified by the formular (26).

R_i: If $y(k)$ is A_i and $u(k)$ is B_i
then $y_P(k + 1) = a_i y(k) + b_i u(k)$, $i = 1, 2, \cdots, K$ (26)

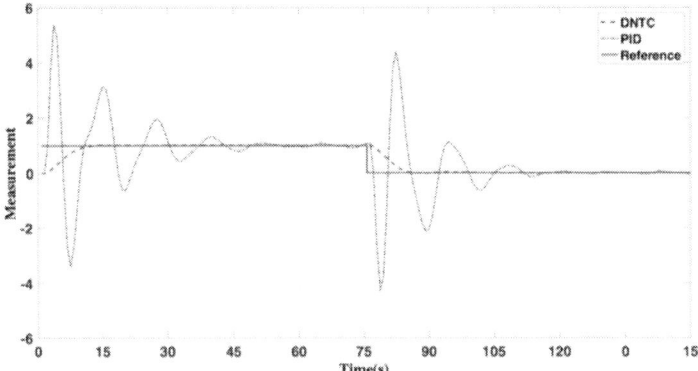

Figure 7. Control performance of DNTC (10-step delay) for OM.

Thus, the predictive model can be derived by

$$y(k+1) = \frac{\sum_{i=1}^{k} \mu_{A_i} \cdot \mu_{B_i}(a_i y(k) + b_i u(k))}{\sum_{i=1}^{k} \mu_{A_i} \cdot \mu_{B_i}}$$

$$= \frac{\sum_{i=1}^{k}(Net_{A_i} \cdot y(k))(Net_{B_i} \cdot u(k))(a_i y(k) + b_i u(k))}{\sum_{i=1}^{k}(Net_{A_i} \cdot y(k))(Net_{B_i} \cdot u(k))} \quad (27)$$

$$= \frac{\sum_{i=1}^{k} Net_{A_i} \cdot Net_{B_i}(a_i y(k) + b_i u(k))}{\sum_{i=1}^{k} Net_{A_i} \cdot Net_{B_i}}$$

Then we can get the one step predictive model.

$$y_p(k+1) = \tilde{a}_i y(k) + \tilde{b}_i u(k) \quad (28)$$

Where $\tilde{a}_i = \dfrac{\left(\sum_{i=1}^{k} Net_{A_i} \cdot Net_{B_i} \cdot a_i\right)}{\left(\sum_{i=1}^{k} Net_{A_i} \cdot Net_{B_i}\right)}$ $\tilde{b}_i = \dfrac{\left(\sum_{i=1}^{k} Net_{A_i} \cdot Net_{B_i} \cdot b_i\right)}{\left(\sum_{i=1}^{k} Net_{A_i} \cdot Net_{B_i}\right)}$

$y_p(k+2)$, $y_p(k+3)$, ☐☐☐ $y_p(k+P)$, where P is the predictive step, $0 \leq P$, can be obtained by recurrence.

$$y_p(k+2) = \tilde{a}_i y(k+1) + \tilde{b}_i u(k+1)$$
$$y_p(k+3) = \tilde{a}_i y(k+2) + \tilde{b}_i u(k+2)$$
$$\vdots$$
$$y_p(k+P) = \tilde{a}_i y(k+P-1) + \tilde{b}_i u(k+P-1) \qquad (29)$$

To increase the prediction accuracy, predicted values at sample time $k+1$, $k+2$, ···, $k+P$ are corrected by the difference between actual process output $y(k)$ and the predicted output $y_p(k)$, i.e.,

$$y_p(k+2) = \tilde{a}_i y(k+1) + \tilde{b}_i u(k+1) + \alpha(y(k) - y_p(k))$$
$$y_p(k+3) = \tilde{a}_i y(k+2) + \tilde{b}_i u(k+2) + \alpha(y(k) - y_p(k))$$
$$\vdots$$
$$y_p(k+P) = \tilde{a}_i y(k+P-1) + \tilde{b}_i u(k+P-1) + \alpha(y(k) - y_p(k)) \qquad (30)$$

where $\alpha \leq 1$ is corrective coefficient, the bigger the α is, the stronger the corrective action will be. To simplify the design and integrate with the local control, we let the Prediction Horizon (PH) be equal to the Control Horizon (CH), $PH = CH = P + 1$. The controller can be designed in the same way. Please reference (14)-(19).

To validate the accuracy of the simplified model, let us take the system (20) as an example. The three individual local models (21) can be replaced by

R1 : If $y(k)$ is A_1 and $u(k)$ is B_1
then $y(k+1) = 0.9962y(k) + 3.3785u(k)$ (31)
R2 : If $y(k)$ is A_2 and $u(k)$ is B_2
then $y(k+1) = 0.9958y(k) + 3.7247u(k)$
R3 : If $y(k)$ is A_3 and $u(k)$ is B_3
then $y(k+1) = 0.9958y(k) + 3.9998u(k)$

Thus, the basic predictive model is obtained.

$$y_p(k+1) = 0.996y(k) + 3.6199\, u(k) \qquad (32)$$

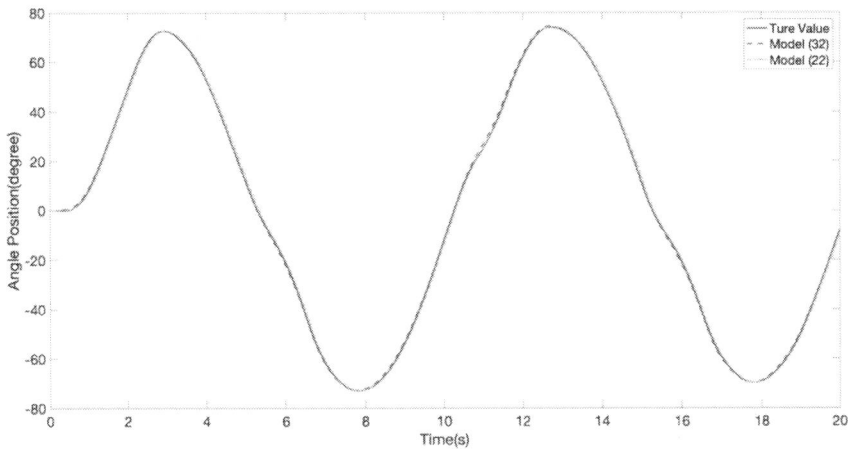

Figure 8. Accuracy comparison of the predictive models.

Figure 8 shows the comparing results of the model outputs (22) and (32). It can be seen that the simplified model is agreed with the original one.

CONCLUSION

For the networked system with blurred mechanism, this chapter has designed a DNTC structure. It uses FCM technology to model a process only according to the input-output data instead of a predefined model. Errors and change of errors between the desired references and the predicted outputs of the process are regarded as inputs of a FSMC and the outputs of the FSMC are candidate future control actions which are used to actively compensate for the time delay by selecting appropriate control action. Simulations show the good performance of the proposed approach. The outputs of the process agree very well with the desired references.

ACKNOWLEDGMENTS

This work is supported by the Science and Technology Program of Beijing Municipal Education Commission (KM201811417001, KM202011417004), the Beijing Natural Science Foundation-Beijing Municipal Education Commission Joint Fund (KZ201811417048), the Beijing Natural Science Foundation-Rail Transit Joint Fund (L191006) and the open research fund of the State Key Laboratory for Management and Control of Complex Systems (20210111).

REFERENCES

[1] Liu, G. P., Mu, J. X. & Chai, S. C. (2006). Design and stability analysis of networked control systems with random communication time delay using the modified MPC. *International Journal of Control*, 79(4), 288-297.

[2] Zhang, H. & Wang, J. M. (2013). Observer-based tracking controller design for networked predictive control systems with uncertain Markov delays. *Proceedings of American Control Conference*, 86(10), 1824-1836.

[3] Pang, Z. H., Liu, G. P., Zhou, D. H. & Chen, M. Y. (2014). Output tracking control for networked systems: A model-based prediction approach. *IEEE Transactions on Control Systems Technology*, 61(9), 4867- 4877.

[4] Gao, H. J. & Chen, T. W. (2008). Networked-based H∞ output tracking control. *IEEE Transactions on Automatic Control*, 53(9), 2142- 2148.

[5] Wang, Y. L. & Yang, G. H. (2008). Output tracking control for networked control systems with time delay and packet dropout. *International Journal of Control*, 81(11), 1709-1719.

[6] Qiu, L., Shi, Y., Pan, J. F. & Xu, G. (2016). Networked H∞ controller design for a direct-rrive linear motion control system. *IEEE Transactions on Industrial Electronics*, 63(10), 6281-6291.

[7] Chen, Z., Pan, Y. & Gu, J. (2015). Integrated adaptive robust control for multilateral teleoperation systems under arbitrary time

delays. *International Journal of Robust and Nonlinear Control, 26*(12), 2708- 2728.

[8] Tong, S. W., Qian, D. W. & Wang, S. (2016). Data-based networked tracking control system with time delay in the forward channel. *Proceedings of the 35th Chinese Control Conference*, 7440-7444.

[9] Tong, S. W., Qian, D. W. & Liu, G. P. (2014). Networked predictive fuzzy control of systems with forward channel delays based on a linear model predictor. *International Journal of Computers, Communications and Control, 9*(4), 471-481.

[10] Nagarale, R. M. & Patre, B. M. (2016). Exponential function based fuzzy sliding mode control of uncertain nonlinear systems. *International Journal of Dynamics and Control, 4*(1), 67-75.

[11] Ngo, Q. H., Nguyen, N. P. & Chi, N. N. (2015). Fuzzy sliding mode control of container cranes. *IET Control Theory and Applications, 13*(2), 419-425.

[12] Hu, H., Liu, G. P. & Rees, D. (2008). Networked predictive control over the internet using round-trip delay measurement. *IEEE Transactions on Instrumentation and Measurement, 57*(10), 2231-2241.

[13] Tong, S. W., Qian, D. W. & Fang, J. J. (2015). Sliding mode output tracking control based on a fuzzy clustering model. *Proceedings of the 2015 International Conference on Advanced Mechatronic Systems*, 228-232.

[14] Gustafson, D. E. & Kessel, W. C. (1979). Fuzzy clustering with a fuzzy covariance matrix. *IEEE Conference on Decision and Control including the 17th Symposium on Adaptive Processes*, 761-766.

[15] Liu, G. P., Rees, D., Chai, S. C. & Nie, X. Y. (2005). Design, simulation and implementation of networked predictive control systems. *Measurement and Control, 38*, 17-21.

[16] Gu, G. Y., Zhu, L. M., Su, C. Y. & Ding, H. (2013). Motion control of piezoelectric positioning stages: modeling, controller design, and experimental evaluation. *IEEE/ASME Transactions on Mechatronics, 18*(5), 1459-1471.

[17] Sun, W. C., Tang, S. W., Gao, H. & Zhao, J. (2016). Two time-scale tracking control of nonholonomic wheeled mobile robots.

IEEE Transactions on Control Systems Technology, 24(6), 2059-2069.

[18] Cao, R. Z. & Low, K. S. (2009). A repetitive model predictive control approach for precision tracking of a linear motion system. *IEEE Transactions on Industrial Electronics, 56*(6), 1955-1962.

[19] Chen, Z., Yao, B. & Wang, Q. (2015). μ-synthesis-based adaptive robust control of linear motor driven stages with high-frequency dynamics: A case study. *IEEE/ASME Transactions on Mechatronics, 20*(3), 1482-1490.

[20] Sun, W. C., Zhang, Y. F. & Huang, Y. P. (2016). Transient-performance guaranteed robust adaptive control and its application to precision motion control systems. *IEEE Transactions on Industrial Electronics, 63*(10), 6510-6518.

[21] Yao, J. Y., Jiao, Z. X. & Ma, D. W. (2014). RISE-based precision motion control of DC motors with continuous friction compensation. *IEEE Transactions on Industrial Electronics, 61*(12), 7067-7075.

[22] Liu, G. P., Chai, S. C. & Rees, D. (2006). Networked predictive control of internet/intranet based system. *Preceedings of the 25th Chinese Control Conference*, 2025-2029.

[23] Li, H. Y., Wu, C. W., Jing, X. J. & Wu, L. J. (2016). Fuzzy tracking control for nonlinear networked systems. *IEEE Transactions on Cybernetics, 47*(8), 2020-2031.

In: Networked Control Systems
Editors: S. Tong and D. Qian

ISBN: 978-1-53619-892-8
© 2021 Nova Science Publishers, Inc.

Chapter 3

Design and Implementation of a Data-Based Adaptative Networked Tracking Control System with NetCon

Shiwen Tong[1,3]* and Ye Zhao[2]

[1]College of Robotics, Beijing Union University, Beijing, China
[2]College of Urban Rail Transit and Logisitics, Beijing Union University, Beijing, China
[3]State Key Laboratory for Management and Control of Complex Systems, Institute of Automation, Chinese Academy of Sciences, Beijing, China

Abstract

The time variation of input signal, time delay and data dropout caused by network and other factors will reduce the performance of networked control, especially the networked tracking control of some processes whose mechanism are unclear, only the input and output data of the system are available. In order to solve this problem, we have proposed an adaptive networked tracking control method based on data and realized it on NetCon system. Firstly, the input-output data are obtained by applying the signal

---

* Corresponding Author's E-mail: shiwen.tong@buu.edu.cn.

excitation to the controlled system, and then the T-S fuzzy model of the controlled object is established by using the fuzzy clustering technology. Then, fuzzy model is transformed into a fuzzy singleton model to obtain fuzzy predictor which can predict process input at future time. In order to realize active compensation, the fuzzy singleton model is transformed into an inverse model according to the reversibility condition, which can generate the future control action. The adaptive control strategy and the internal model structure are adopted to eliminate the external disturbance and the system uncertainty. Simulation results in NetCon system show the effectiveness of the proposed method.

Keywords: inverse model control, adaptative strategy, delay compensation, NetCon system

1. INTRODUCTION

Networked control system is a closed-loop structure connected by network cables among sensors, controllers and actuators. Its system complexity is low, and reliability is high, and it is easy to maintain. However, the inevitable time delay, data dislocation and packet loss in network communication affect the control performance of the system, and the traditional prediction algorithm cannot effectively solve such problems [1]. In recent years, experts and scholars have proposed a lot of control methods. In order to reduce the influence of sensor data loss in the control feedback loop, Tefili et al. designed a fuzzy controller, which compensated the data loss through zero input and hold input, and used genetic algorithm to minimize the difference between the actual output and the required output. Compared with PID, it showed better control performance [2]. Fan et al. evaluated the influence of time delay on NCS by studying the time delay boundary, and proposed a two-way scheduling algorithm for distributed networked control of different types of transmission packets, and the effectiveness of the mechanism was verified by experiments [3]. Yan et al. constructed the sliding mode controller to drive the system state trajectory to the sliding mode surface designed in advance, and designed the sliding mode controller of the networked control system [4]. Lu et al. designed a networked control system jointly controlled by remote and local controllers. After receiving data through an unreliable communication channel, the remote controller sends confirmation information to the local controller, and the two controllers respond and send the final information to the plant [5]. Li et al. proposed finite-time state-feedback and output-feedback predictive controllers for data misalignment caused by network in feedback channels and forward channels, and provided sufficient conditions for finite time stability [6]. Yang et al. introduced the event triggering mechanism, estimated the system state through the received output measurements, and developed a model-based network predictive control scheme. The extended model and piecewise linear model were adopted to provide sufficient conditions to ensure the system performance [7]. Mohareri et al. proposed an adaptive trajectory tracking controller, and the simulation and experiment on the mobile robot platform verified the performance of the control algorithm [8].

To sum up, most control methods are designed based on the model and need to know the model parameters and other information. Although some researchers have proposed prediction ideas [9-12], most of them are passive compensation methods for network delay. Therefore, this chapter proposes a data-based adaptive networked tracking control method (DANTC). Taking NetCon system as the platform, the design of predictor and inverse model controller is designed in detail, and the network real-time simulation is carried out with the servo control system as the controlled object.

2. DESIGN OF DANTC IN NETCON SYSTEM

2.1. Core Idea of DANTC

The structure of DANTC in NetCon system is shown in Figure 1. The upper configuration part mainly includes the controller side and controlled object side. Firstly, the input and output data of the system are obtained by signal excitation. Then, the T-S fuzzy model of the system is obtained by constructing the fuzzy clustering model. Considering to the equivalence effect of the system, the T-S fuzzy model is transformed into a fuzzy singleton model. So we get the fuzzy model predictor. It can predict the future output of the controlled object. According to the reversibility condition, the fuzzy singleton model is transformed into the fuzzy inverse model, which can generate the control action at the future time [13]. The predicted control action is sent from the controller to the controlled object through the network transmitter to realize active compensation. The process side selects the optimal control action at different times from the received data packets [14]. In order to eliminate external disturbance and system uncertainty, an internal model control structure and an adaptive strategy are adopted, and the parameters of the inverse model controller and the predictor are updated by adding a feedback correction module, which can improve the control accuracy.

Through the network cable connection, the upper configuration algorithm structure (IP address: 192.168.0.151) can be downloaded to Netcontroller1 (IP address: 192.168.0.151) and Netcontroller2 (IP address: 192.168.0.152), the controller receives the inverse model algorithm, and the sensor receives the prediction algorithm. The servo control system is connected to the sensor through a digital-to-analog connector. The sensor processes the collected data and sends it to the controller. The controller executes the control algorithm to obtain future control actions and feeds them back to the sensor, and finally sends them to the servo control system, thus forming a closed-loop network control. The upper monitoring software (NetConTop) realizes real-time monitoring of signals change in Netcontrollers by means of network connection.

2.2. Predictor Design

Because the system model is not clear, it is necessary to apply the signal excitation to the system to obtain the system input-output data pairs, and then use the fuzzy cluster modeling technology [15] to establish mathematical model of the system. The algorithm of fuzzy cluster modeling (FCM) is shown in Table 1.

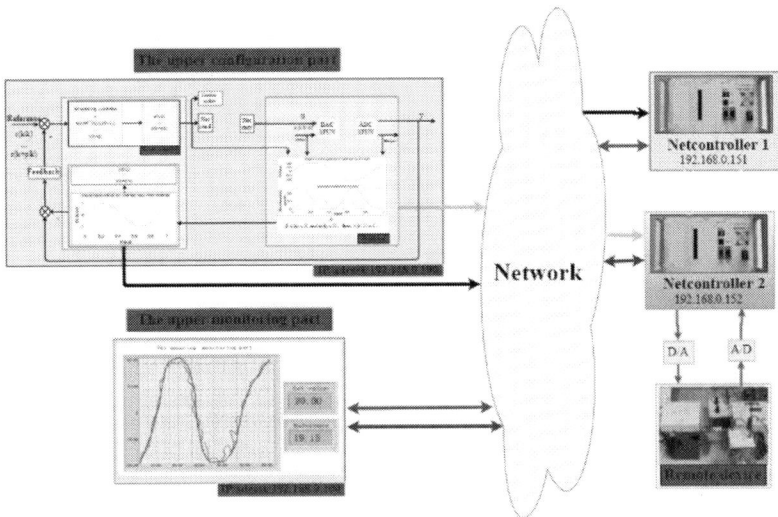

Figure 1. Realization of DANTC structure in NetCon system.

The core of the predictor proposed in this chapter is to predict the output of the controlled object at the future moment. So we firstly use T-S fuzzy model (1) to describe the system.

R_i: If $y(k)$ is A_i, $y(k-1)$ is G_i and $u(k)$ is B_i

then $y(k + 1) = a_i y(k) + g_i y(k-1) + b_i u(k), i = 1, 2, \cdots, K$ (1)

In which, A_i, B_i and G_i are fuzzy sets of the antecedent parts, a_i, b_i, and g_i are parameters of consequent parts. The parameters of the antecedent parts can be obtained by G-K Algorithm [16], and the parameters of the consequent parts can be calculated by least square method.

However, the T-S fuzzy model is a multivariate mapping relation, and the future control action cannot be obtained by inverse calculation. Therefore, the second step of the predictor design is to convert the clustering model into a fuzzy singleton model by using the equivalence (2) of the system. Then $y(k + 1) = f(x(k), u(k))$ will be converted into $y(k + 1) = fx(u(k))$. The detailed algorithm of the predictor is shown in Figure 2.

Table 1. Algorithm of FCM

Algorithm 1: Fuzzy clustering model
Input: Input-output data pairs of system A, A = $\{x_k | k = 1, \ldots, n\}$
Output: Membership matrix U_{ij} j = 1, …, c, j = 1, …, n

Initialization: cluster member c
   fuzzification parameter m
   termination criterion r
   define membership matrix U_{ij}
   define the objective function $J^{(0)}$

While $\|J^{(c+1)} - J^{(c)}\| < r$ do

calculate the center vector a_i
   update membership matrix U_{ij}
   update the objective function

Endwhile

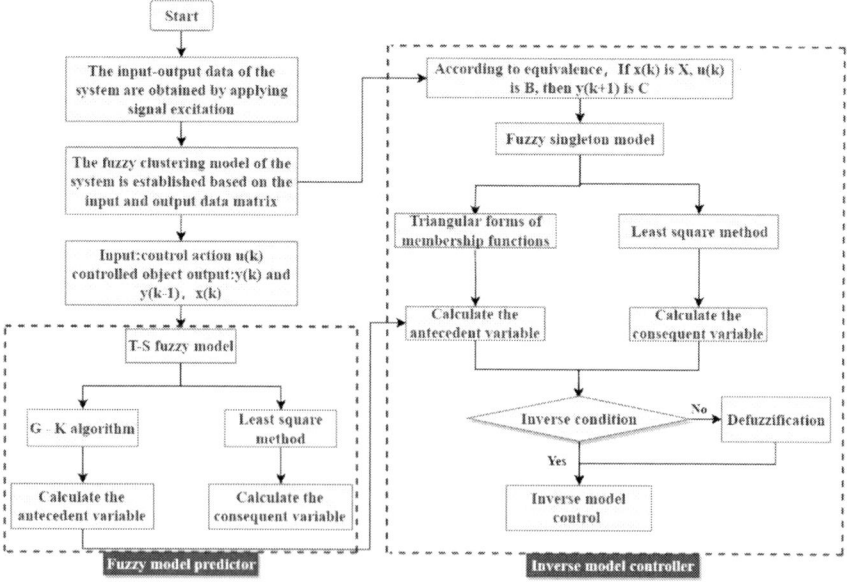

Figure 2. Algorithm of fuzzy predictor.

If x(k) is X, and u(k) is B then y(k + 1) is C   (2)

The y(k + 1), a model output can be derived by equation [17].

$$y(k+1) = \frac{\sum_{i=1}^{M}\sum_{j=1}^{N}\beta_{ij}(k)c_{ij}}{\sum_{i=1}^{M}\sum_{j=1}^{N}\beta_{ij}(k)} = \frac{\sum_{i=1}^{M}\sum_{j=1}^{N}\mu_{X_i(x(k))}\mu_{B_j(u(k))}c_{ij}}{\sum_{i=1}^{M}\sum_{j=1}^{N}\mu_{X_i(x(k))}\mu_{B_j(u(k))}}$$

(3)

Different from the membership function of fuzzy clustering model, the membership function in triangular form of fuzzy singleton model has at least one more partition than (1), and the intersect points of adjacent membership function of system (1) become the core of triangular membership function in (2). The solving process of triangular membership function of system (2) is shown in Table 2.

Table 2. Algorithm of triangular membership functions

Algorithm 2: Triangular membership functions
Input: Control action u(k), controlled object output y(k) and y(k-1)
Output: Triangular membership functions µ

Initialization: sample time t
  fuzzy cluster model MiuA$_i$, i = 1, ..., n
  fuzzy singleton model MiuB$_i$, i = 1, ..., n
  input x(k$_i$)
  consequent parameters: a$_i$, b$_i$, i = 1, ..., n

For T = 1: V% control actions input are V vectors;

u(k) is current moment u(T);
y(k) = u(1 + T)
if y(k) < a$_i$: calculate MiuA$_i$ end
if y(k) < = a$_{i+1}$ and y(k) > = a$_{i+2}$: calculate MiuA$_{i+1}$ end
if y(k) > = a$_i$ and y(k) < = a$_{i+3}$: calculate MiuA$_{i+2}$ end
if y(k) > a$_{i+1}$: calculate MiuA$_{i+3}$ end
if u(k) < b$_i$: calculate MiuB$_i$ end
if u(k) < = b$_{i+1}$ and u(k) > = b$_{i+2}$: calculate MiuB$_{i+1}$ end
if u(k) > = b$_i$ and u(k) < = b$_{i+3}$: calculate MiuB$_{i+2}$ end
if u(k) > b$_{i+1}$: calculate MiuB$_{i+3}$ end
for i = 1:n:
  for j = 1:n:
    Beita$_{ij}$ = MiuA$_i$*MiuB$_j$
end
  end

End

Table 3. Algorithm of inverse model controller

Algorithm 3: Inverse Model Controller
Input: Reference signal r(k), difference e of predictor output y(k + 1) and controlled object output y(k)
Output: Control action u(k)

Initialization: r(k + 1), u(k), x(k),
    triangular membership function MiuA$_i$, MiuC$_j$, i, j = 1 ,..., n
    consequent parameters: a$_i$, b$_i$, i = 1, ..., n

For T = 1:V;

rk1 = u(T);
 xk = u(T + 12);
 y(k) = xk
 if y(k) < a$_i$: calculate MiuA$_i$ end
 if y(k) < = a$_{i+1}$ and y(k) > = a$_{i+2}$: calculate MiuA$_{i+1}$ end
 if y(k)k > a$_i$ and y(k) < = a$_{i+3}$: calculate MiuA$_{i+2}$ end
 if y(k) > a$_{i+1}$: calculate MiuA$_{i+3}$ end
 tmpc = u(:);
 c$_{ij}$ = reshape(tmpc, n, n)
 for h = 1:n:
    c$_{jh}$ = [MiuA$_1$...MiuA$_n$]* c$_{ij}$(:,h)
 end
 if rk1 < c$_{j2}$: calculate MiuC$_i$ end
 if rk1 < = c$_{j3}$ and rk1 > = c$_{j1}$: calculate MiuC$_{i+1}$ end
 if rk1 > = c$_{j2}$ and rk1 < = c$_{j4}$: calculate MiuC$_{i+2}$ end
 if rk1 > c$_{j4}$: calculate MiuC$_{i+3}$ end
 for i = 1:n:
 u(k) = sum(MiuC$_i$*b$_i$)
 end

End

After the triangular membership is obtained through the above algorithm, y(k + 1) can be calculated according to (2.3). However, y(k + 1) = f(x(k), u(k)) which cannot directly generate control actions. Therefore, u(k) would be obtained through inversion by inverse model controller.

2.3. Inverse-Model-Based Controller Design

If the fuzzy singleton model in the predictor meets the reversibility condition, the inverse model controller can be obtained by inverting the model, and then the control action u(k) of the controlled object at the future time would be obtained. Control actions are sent to the controlled object side through the network transmitter to compensate the delay in the network. In order to improve the control accuracy, the difference vector between the actual output and the expected output of the controlled object is used as a feedback signal to flow into the controller, and the parameters of the antecedent and consequent parts in the controller and the predictor are dynamically updated. The fuzzy singleton model in the predictor is invertible if and only if the following conditions [17, 18] are satisfied.

Define b_i is core of B_i:

For each B_j, the core b_j is single point, that is $|B_j| = 1$, $j = 1, ..., N$ and

$$b_1 < ... < b_N \ c_{i1} < ... < c_{iN} \text{ or } c_{i1} > ... > c_{iN}, \ i = 1, ..., M \tag{4}$$

The relation between the output u(k) of the inverse model controller and the output y(k + 1) of the fuzzy model predictor is as following Equation (5). The networked control scheme proposed in this chapter only considers the delay in the forward channel and ignores the delay in the backward channel.

$$u(k) = f_x^{-1}(y(k+1)) \tag{5}$$

According to [15, 21], we can see that if there is a d-order delay in the forward channel, then y(k + 1) = r(k + d + 1). And let r(k + d + 1) = r(k + 1), then y(k + 1) = r(k + 1). So the Equation (6) would be derived.

$$u(k+1) = f_x^{-1}(r(k+2))$$
$$...$$
$$u(k+p) = f_x^{-1}(r(k+p+1)) \tag{6}$$

The algorithm structure of the inverse model controller is shown in Table 3.

2.4. Delay Compensation Design

This chapter classifies data misalignment and data packet loss into network delay, and sets the control time domain to be no less than the forward channel delay [19]. Since the network communication channel can transmit data packets, we can send the future control actions u(k), u(k + 1), ..., u(k + p) which respond to sampling time k to the controlled object side through the network. And the controlled process selects the optimal control action from the received sequence to compensate for the time delay. Assume that the latest control sequence received by the controlled object is:

$$\begin{bmatrix} u(k-t \mid k-t) \\ u(k-t+1 \mid k-t) \\ \vdots \\ u(k-t+P \mid k-t) \end{bmatrix} \quad (7)$$

The control action selected by the controlled object end is:

$$u(k) = u(k \mid k - t) \quad (8)$$

As for the case of data packet loss, it can be regarded as a special case of the time delay problem, that is, the case of infinite time delay. If the delay is infinite, no data is received and the data can be considered lost. Therefore, if the delay has an upper bound t_m, $u(k|k-t_m)$ can be used to replace the impact of data loss on system performance if data loss occurs.

3. RELIZATION IN NETCON SYSTEM

In order to verify the function of the DANTC algorithm in NetCon system, the discrete system represented by the servo motor system (9) is selected as the controlled object to verify the feasibility of the algorithm [20].

$$G(z^{-1}) = \frac{-0.0086z^{-1} + 1.268227z^{-2}}{1 - 1.66168z^{-1} + 0.6631z^{-2}} \qquad (9)$$

where sample time is 0.04s.

In this experiment, two Netcontrollers are used as controller and sensor respectively. The program block diagram shown in Figure 3 is divided into two parts. One part is the inverse model controller, which can generate the control action u(k + d) through the signal input r(k + d). Due to the network delay, the action received by the sensor part in the block diagram is u(k). The other part is the fuzzy model predictor, which can predict the output y(k + 1) of the controlled object at the future moment. The forward channel is connected by network transmitter and receiver, and the backward channel is wired for signal transmission. One of the features is adaptive control, which can dynamically update the parameters in inverse model and fuzzy model through feedback correction of the difference vector between the actual output and the expected output of the controlled object. The block diagrams are compiled and downloaded to controller (Netcontroller1, IP address 192.168.0.151) and sensor (Netcontroller2, IP address 192.168.0.152).

Firstly, three sine signals with the frequencies of 0.012 π, 0.03 π and 0.06 π rad/sec and the amplitude of 0.4 are selected to incentive the system. Then three clusters are selected, the fuzzification parameter m is determined as 2, and the termination criterion of the program equals 0.01. Then the prediction simplified models (10) is obtained according to GK algorithm.

$$y_p(k + 1) = 1.6987y(k) - 0.7000y(k-1) + 1.0844u(k) \qquad (10)$$

3.1. Case 1

In this situation, considering the 10-step delay in the forward channel, the sine signal with frequency of 0.2 π and amplitude of 20 is selected as the desired signal to compare the control performance of DANTC algorithm and PID controller (11). Where P = 0.008, I = 0.003, D = 0.00001 and Ts = 0.04.

$$P + I \cdot T_s \frac{1}{z-1} + D \cdot \frac{1}{T_s} \frac{z-1}{z} \tag{11}$$

As shown in Figure 4, the red solid line represents the expected signal, the green dotted line represents the predicted output of the DANTC algorithm, and the blue solid line represents the output of the PID controller. It can be seen from the figure that the algorithm proposed in this chapter presents a relatively high precision and fast dynamic response when processing the forward communication channel delay.

3.2. Case 2

This case is used to test the control effect of DANTC algorithm in the case of packet loss in the network.

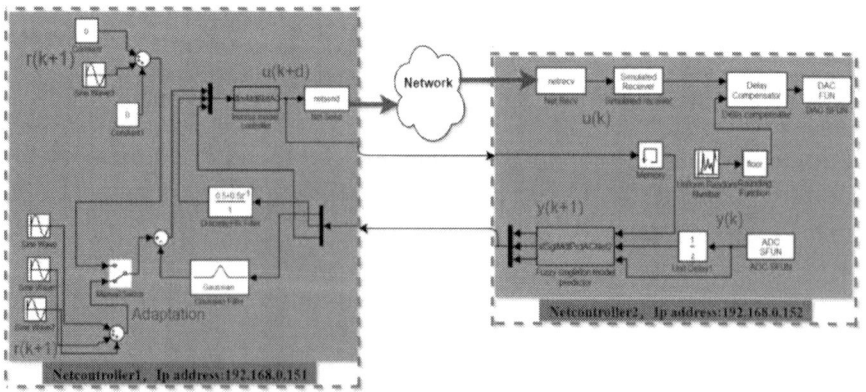

Figure 3. Simulink digram of DANTC in NetCon system.

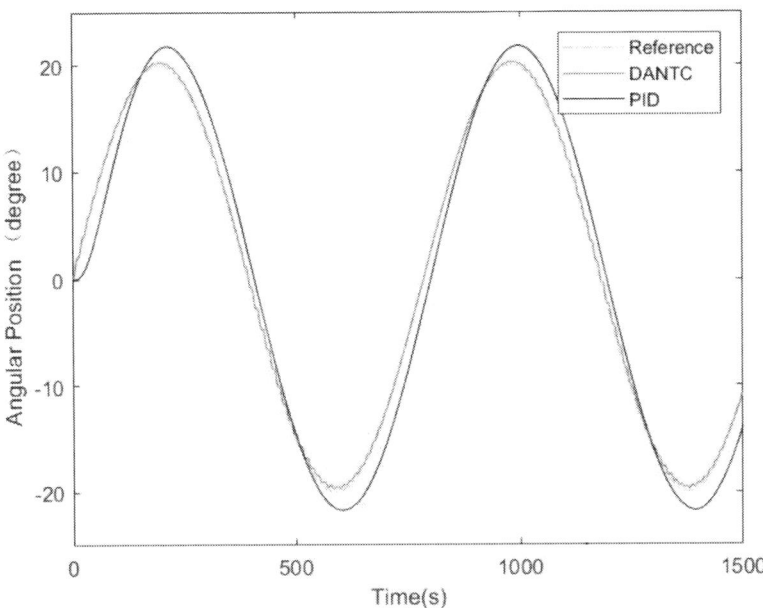

Figure 4. Control performance comparison with 10-step delay.

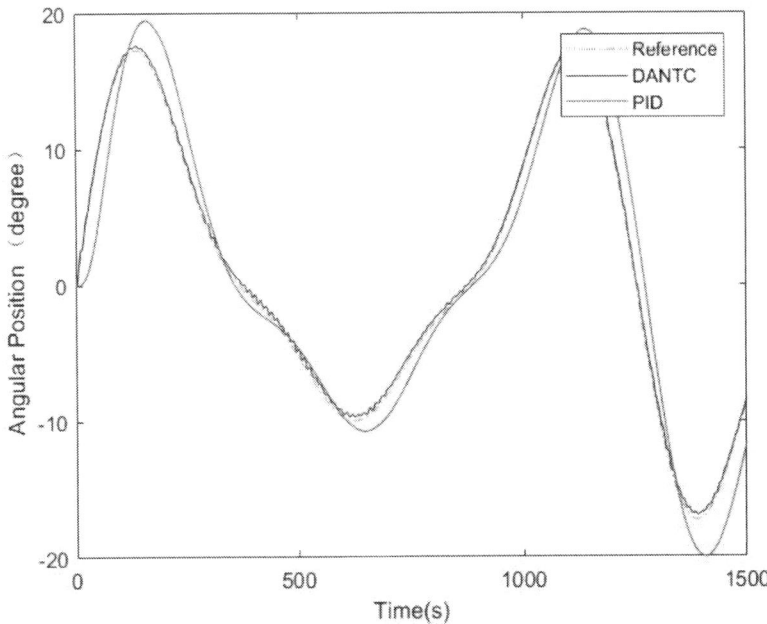

Figure 5. Control performance comparison with data packet dropout.

It is assumed that the system has a random time delay of 15 steps, and the delay between 11-15 steps is used to simulate the situation of data packet loss. We select the sine signal with frequencies of 0.1 π, 0.14 π, 0.06 π and amplitude of 20 to compare the control performance of DANTC algorithm. The PID parameter setting is the same as case 1.

In Figure 5, we can see that compared with PID control, DANTC shows the performance that the reference signal and the output of the controlled object match very well.

CONCLUSION

1. For the networked control of time-varying signals with fuzzy mechanism and incomplete model, the data-based adaptive tracking control method proposed in this chapter can be used. And we have realized it in NetCon system.
2. The core idea of this method is to design a fuzzy model predictor and inverse-model-based controller to obtain future control actions to actively compensate time delay existing in the network.
3. Internal model structure and adaptative strategy are adopted to implement feedback correction which can improve prediction accuracy.

ACKNOWLEDGMENTS

This work is supported by the Science and Technology Program of Beijing Municipal Education Commission (KM201811417001, KM202011417004), the Beijing Natural Science Foundation-Beijing Municipal Education Commission Joint Fund (KZ201811417048), the Beijing Natural Science Foundation-Rail Transit Joint Fund (L191006).

REFERENCES

[1] Li, P. F. (2020). *Research on model predictive control theory and method in networked systems*. University of Science and Technology of China.

[2] Tefili, D., Aoki, A. R., Leandro, G. V., Ribeiro, E. P. (2021). Performance improvement for networked control system with nonlinear control action. *International Journal of Dynamics and Control*.

[3] Fan, Z. M., Yu, X. J., Wan, H., Kang, M. L., Liu, Y., He, Y., Yazdani, G. X. S. (2020). A time-delay-bounded data scheduling algorithm for delay reduction in distributed networked control systems. *Mathematical Problems in Engineering*.

[4] Yan, C. J., Ding, S. F., Yang, R. N. (2020). Sliding mode control for discrete-time networked control systems based on the NPC scheme. *IEEE 16th International Conference on Control and Automation*, 536-540.

[5] Lu, X., Xu, J. J., Liang, X. (2020). Stabilisation of networked control systems with remote and local controllers subject to delay and packet dropout. *IET Control Theory and Applications*, 14(14), 2008-2015.

[6] Li, Y. J., Liu, G. P., Sun, S. L., Tan, C. (2019). Prediction-based approach to finite-time stabilization of networked control systems with time delays and data packet dropouts. *Neurocomputing*, 329, 320-328.

[7] Yang, R. N., Zheng, W. X. (2020). Output-based event-triggered predictive control for networked control systems. *IEEE Transactions on Industrial Electronics*, 67(12), 10631-10640.

[8] Mohareri, O., Dhaouadi, R., Rad, A. B. (2012). Indirect adaptive tracking control of a nonholonomic mobile robot via neural networks. *Neurocomputing*, 88, 54-66.

[9] Jia, X. C., Zhang, D. W., Hao, X. H. (2009). Fuzzy H tracking control for nonlinear networked control systems in T-S fuzzy model. *IEEE Transactions on Systems, Man, and Cybernetics*, 39, 1073-1079.

[10] Zheng, Y., Fang, H. J., Wang, H. O. (2006). Takagi-Sugeno fuzzy-model-based fault detection for networked control systems

with markov delays. *IEEE Transactions on Systems, Man, and Cybernetics,* 36(4), 924-929.

[11] Mahmoud, M. S. (2012). H control of uncertain fuzzy networked control systems with state quantization. *Intelligent Control and Automation*, 3, 59-70.

[12] Li, Y. H., Li, Y. L. (2017). Robust L1 output tracking control for uncertain networked control systems described by T-S fuzzy model with distributed delays. *International Journal of Systems Science*, 48(15), 3296-3304.

[13] Tong, S. W., Qian, D. W., Liu, G. P. (2015). Networked predictive fuzzy control of systems with forward channel delays based on a linear model predictor. *Int. J. Comput. Commun. Control*, 9(4), 228-232.

[14] Tong, S. W., Qian, D. W., Fang, J. J. (2015). Sliding mode output tracking control based on a fuzzy clustering model. *Proceedings of the 2015 International Conference on Advanced Mechatronic Systems*, 228-232.

[15] Tong, S. W., Oian, D. W., Chen, G. (2018). Design and simulation of an adaptive networked tracking control system. *International Conference on Advanced Mechatronic Systems*, 66-71.

[16] W. S. Hu, G. P. Liu, D. Rees (2008). Networked predictive control over the internet using round-trip delay measurement. *IEEE Transactions on Instrumentation and Measurement,* 57(10), 2231-2241.

[17] D. E. Gustafson, W. C. Kessel (1978). Fuzzy clustering with a fuzzy covariance matrix. *IEEE conference on decision and control*, 761-766.

[18] R. Babuska (1998). Fuzzy Modeling for Control. *Kluwer Academic Publishers.*

[19] Z. H. Pang, G. P. Liu, D. H. Zhou, M. Y. Chen (2014). Output tracking control for networked systems: a model-based prediction approach. *IEEE Transactions on Industrial Electronics*, 61(9), 4867-4877.

[20] X. Y. Ding, G. Shen, X. Li, Y. Tang (2020). Delay compensation position tracking control of electro-hydraulic servo systems based on a delay observer. *Proceedings of the Institution of Mechanical Engineers. Part I: Journal of Systems and Control Engineering*, 234(5), 622-633.

[21] G. P. Liu, J. X. M, R. D. (2006). Design and Stability Analysis of Networked Control Systems with Random Communication Time Delay Using the Modified MPC. *International Journal of Control*, 79(4), 288-297.

In: Networked Control Systems
Editors: S. Tong and D. Qian
ISBN: 978-1-53619-892-8
© 2021 Nova Science Publishers, Inc.

Chapter 4

PID TEMPERATURE CONTROL FOR AN AIR TANK SYSTEM WITH PARAMETERS TUNING THROUGH NETWORK

Shiwen Tong[1,3]* *and Dianwei Qian*[2]

[1]College of Robotics, Beijing Union University, Beijing, China
[2]School of Control and Computer Engineering,
North China Electric Power University, Beijing, China
[3]State Key Laboratory for Management and Control of Complex Systems, Institute of Automation, Chinese Academy of Sciences, Beijing, China

ABSTRACT

This chapter presents an application of Networked Control in the air tank temperature system. The system is controlled by Proportional-Integral-Differential (PID) algorithm running in the Siemens S7-200 PLC. Through an OPC server component, controller parameters (K_p, K_i, K_d) can be remote tuning by Matlab. Thus, complex control algorithms such as fuzzy inference, expert system and genetic optimization can be utilized. The process is supervised by configuration software, for example King Views, located in different geographical areas at the same time.

---

* Corresponding Author's E-mail: shiwen.tong@buu.edu.cn.

Keywords: networked control, OPC interface, PID tuning, PLC, remote supervisory

1. INTRODUCTION

The development of the network has changed our life. Almost anything can be connected to network. We are buying goods from network, watching TV through internet, steering automobiles by CAN bus. NCS (Networked Control System) is formed by combining network with control. This research direction has boundless prospects. Web-based experiments have become feasible and popular in the process of education and research [1-8]. For example, the Italy Siena University has developed an Automatic Control Telelab [4]. Garcia-Zubia etc., have proposed a remote control lab with four-layer distributed structure called Weblab-Deusto [5]. The institute of automation, Chinese Academy of Sciences and the University of South Wales of UK have put forward a networked control lab named NCSLab which has realized remote experiment between China and UK [6-8].

In the chapter, we design a control and supervisory system for an air tank with industrial control products. By using fuzzy inference, PID controller parameters can be remotely tuned through network.

This chapter is organized as follows. In Section 2, an overview of the air tank temperature control system is introduced. Section 3 mainly presents the design of PID controller and the parameters tuning with fuzzy inference through network. Section 4 gives the simulative and experimental results. Finally, the conclusion is drawn.

2. AIR TANK TEMPERATURE CONTROL SYSTEM

The air tank temperature control system (as shown in Figure 1) is divided into two layers. The bottom layer mainly consists of field instruments. Air is pumped into the air tank and heated by an electrical heater through a voltage regulator module, and then blew out of the tank from the top outlet. The temperature of the air tank is controlled by Siemens S7-200 PLC. The flow of the air is regulated by a frequency transformer with panel operation mode. When the flow of the air is constant, the temperature within the air tank can be only regarded as influenced by the electrical heater. The top layer is aimed at process supervisory and controller parameters regulation through network. These two layers are connected by the OPC interface such that the King View and Matlab are as the OPC client and the Siemens PC Access is as the OPC server.

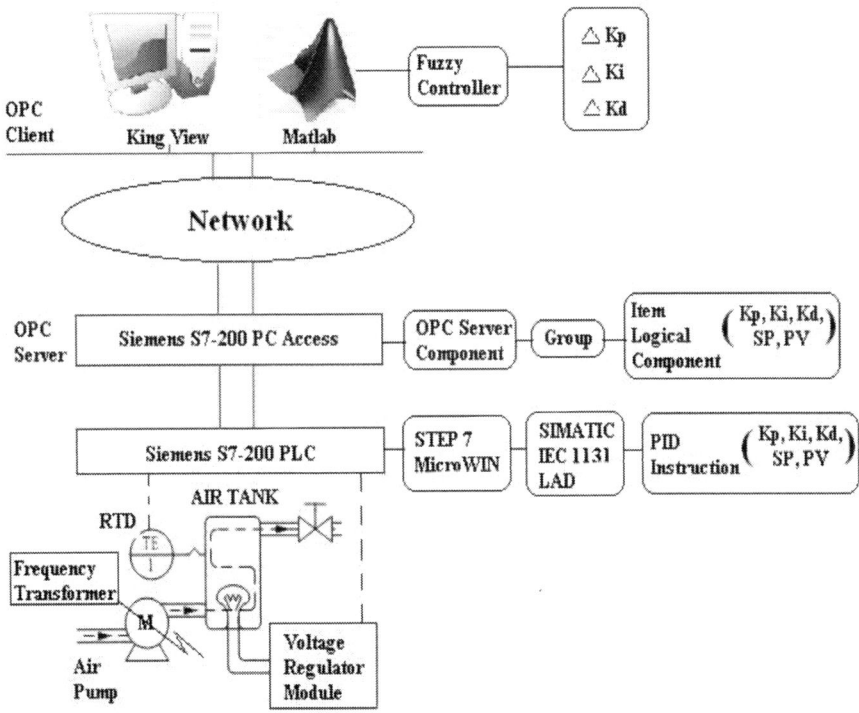

Figure 1. The air tank temperature control system.

3. PID CONTROLLER TUNING THROUGH NETWORK

The air tank temperature system is adopted PID controller with the form

$$u(t) = K_p \cdot e(t) + K_i \cdot \int_0^t e(t)dt + K_d \cdot \frac{de(t)}{dt} \tag{1}$$

where deviation $e(t) = r(t)-y(t)$, $r(t)$ is the set-point, $y(t)$ is the process output, $u(t)$ is the control action, K_p, K_i, K_d are the proportional, integral and differential coefficient, respectively.

Different combinations of parameters K_p, K_i, K_d determine stability, dynamic response and steady-state error.

- Proportional coefficient K_p is used to accelerate dynamic response and improve regulation accuracy of the controlled system. The bigger the proportional coefficient K_p is, the faster the dynamic response will be. The larger the K_p is, the higher the regulation accuracy will be. However, great K_p will cause overshot, even make the system unstable. In contrast, when the value of K_p is too small, the regulation accuracy will be reduced and the dynamic response will be slow. Thus the static and dynamic characteristic of the system will be worse.
- Integral coefficient K_i is used to eliminate the steady-state error of the system. The bigger the integral coefficient K_i is, the faster the static error of the system is eliminated. But great K_i will cause integration saturation phenomenon which will lead to large overshot. On the contrary, if K_i is too small, it will be difficult to eliminate static error.
- Differential coefficient K_d is used to improve the dynamic characteristic of the system. It predicts the varies of the deviation and suppresses the change of deviation in any direction. Great K_d will make the dynamic response brake in advance thus prolonging the regulation time.

From the above analysis it can be seen that the steady-state and the dynamic performance of control systems are determined by parameters K_p, K_i, K_d. While the error and error change between set-points and process values to some extents reflects the steady-state and dynamic performance of the system. Thus one can design a fuzzy controller to regulate parameters K_p, K_i, K_d according to the error and error change between set-points and process values. The parameters tuning structure is shown in Figure 2. The parameters K_p, K_i, K_d of the PID controller are collected through a network. To improve the control performance, incremental values of K_p, K_i, K_d are generated through fuzzy inference with error and error change between set-points and process outputs as inputs. K_p, K_i, K_d are updated by adding incremental values to the last parameters of the PID controller. Then they are sent to the local PID controller via network. Therefore, the control performance has been improved in this way. Three two-input-one-output fuzzy controllers have been designed. It uses the error E and error change EC as the input linguistic variables and incremental K_p, K_i, K_d as output linguistic variables.

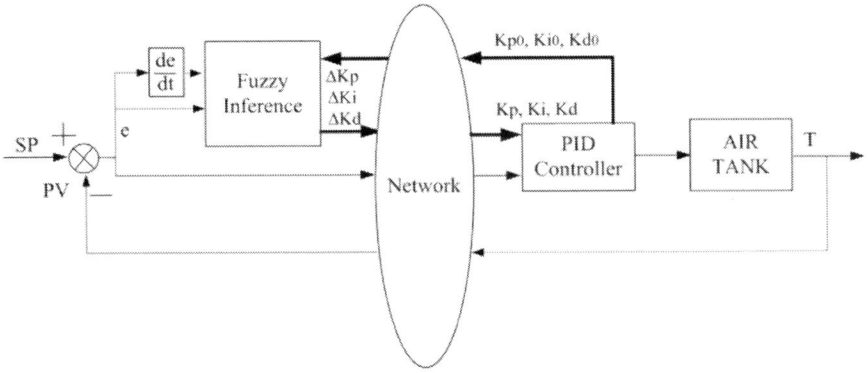

Figure 2. Fuzzy PID with parameters tuning through network.

The normalized domains of error E and error EC are defined as [-2, +2]. They are divided into seven fuzzy subsets with linguistic variables $L(e)/L(ec)$ = {NB, NM, NS, ZE, PS, PM, PB}. The membership functions of E and EC are presented in Figure 3, and the membership functions of $\triangle K_p$, $\triangle K_i$, $\triangle K_d$ are shown in Table 1. They adopt discrete form of membership functions. The normalized domains of $\triangle K_p$, $\triangle K_i$, $\triangle K_d$ are defined as [-3, +3]. They have seven linguistic variables $L(\triangle K_p/\triangle K_i/\triangle K_d)$ = {NB, NM, NS, ZE, PS, PM, PB}.

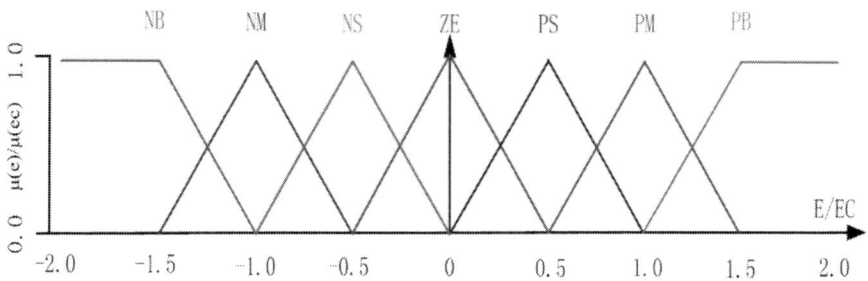

Figure 3. The membership functions of E and EC.

Table 1. Membership functions of variables △K_p, △K_i, △K_d

µ (△K_p)/ µ (△K_i)/ µ (△K_d)	-3	-2	-1	0	+1	+2	+3
NB	1.0	0.7	0.3	0	0	0	0
NM	0.7	1.0	0.7	0.3	0 0.3	0	0
NS	0.3	0.7	1.0	0.7	0.7	0	0
ZE	0	0.3	0.7	1.0	1.0	0.3	0
PS	0	0	0.3	0.7	0.7	0.7	0.3
PM	0	0	0	0.3	0.3	1.0	0.7
PB	0	0	0	0	0	0.7	1.0

From Figure 3, it can be seen that the input variable error is belongs to at most two membership functions, and the input variable error change is also belonging to at most two membership functions. Thus, for every output variable (△K_p, △K_i, △K_d), the maximum of control rules to be fired in the inference process is four. One can focus on these four control rules in each cycle time. A simplified fuzzy inference method is adopted [9, 10]. The control rules for the variables △K_p, △K_i, △K_d are shown as Table 2-Table 4. Take the inference of variable K_p as an example, it can be obtained by the following inference algorithm.

Table 2. Control rules for variable $\triangle K_p$

ec\e	NB	NM	NS	ZE	PS	PM	PB
NB	PB	PB	PM	PM	PS	PS	ZE
NM	PB	PB	PM	PM	PS	ZE	ZE
NS	PM	PM	PM	PS	ZE	NS	NM
ZE	PM	PS	PS	ZE	NS	NM	NM
PS	PS	PS	ZE	NS	NS	NM	NM
PM	ZE	ZE	NS	NM	NM	NM	NB
PB	ZE	NS	NS	NM	NM	NB	NB

Table 3. Control rules for variable $\triangle K_i$

ec\e	NB	NM	NS	ZE	PS	PM	PB
NB	NB	NB	NB	NM	NM	ZE	ZE
NM	NB	NB	NM	NM	NS	ZE	ZE
NS	NM	NM	NS	NS	ZE	PS	PS
ZE	NM	NS	NS	ZE	PS	PS	PM
PS	NS	NS	ZE	PS	PS	PM	PM
PM	ZE	ZE	PS	PM	PM	PM	PB
PB	ZE	ZE	PS	PM	PB	PB	PB

Table 4. Control rules for variable $\triangle K_d$

ec\e	NB	NM	NS	ZE	PS	PM	PB
NB	NB	NB	NM	NM	NS	ZE	ZE
NM	NB	NB	NM	NS	NS	ZE	ZE
NS	NB	NM	NS	NS	ZE	PS	PS
ZE	NM	NM	NS	ZE	PS	PM	PM
PS	NM	PS	ZE	PS	PM	PB	PB
PM	ZE	ZE	PS	PM	PM	PB	PB
PB	ZE	ZE	PS	PM	PM	PB	PB

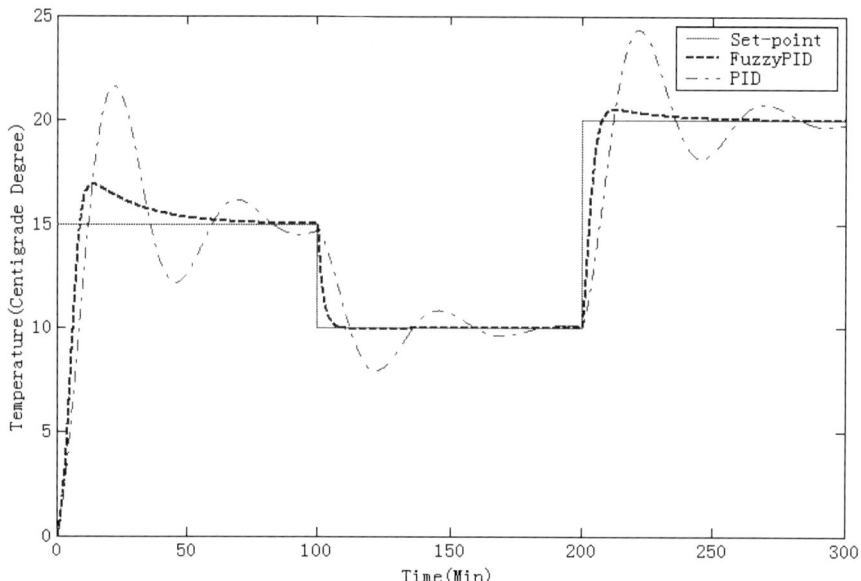

Figure 4. Control performance comparison between PID and Fuzzy PID.

Algorithm 1: Inference of variable K_p
Input: Error between set-point and process value; Error change
Output: Proportional coefficient K_p
Process:

01. Get error and error change according to the set-points and the process values.

02. Map the inputs error and error change to the normalized domains of *E* and *EC* by multiplying the scaling gains with them.

03. Calculate membership functions of *E* and *EC*. Suppose membership functions belong to *E* are μ_{Ai} and μ_{Ai+1}, membership functions belong to *EC* are μ_{Bj} and μ_{Bj+1}, then four control rules $L_{\Delta u}^{(i,j)}, L_{\Delta u}^{(i,j+1)}, L_{\Delta u}^{(i+1,j)}$ and $L_{\Delta u}^{(i+1,j+1)}$ are fired.

04. Get four "If-Then" rules

If $(e = L_e^{(i)}) \cup (ec = L_{ec}^{(j)})$ then $(\Delta K_p = L_{\Delta K_p}^{(i,j)})$

If $(e = L_e^{(i)}) \cup (ec = L_{ec}^{(j+1)})$ then $(\Delta K_p = L_{\Delta K_p}^{(i,j+1)})$

If $(e = L_e^{(i+1)}) \cup (ec = L_{ec}^{(j)})$ then $(\Delta K_p = L_{\Delta K_p}^{(i+1,j)})$

If $(e = L_e^{(i+1)}) \cup (ec = L_{ec}^{(j+1)})$ then $(\Delta K_p = L_{\Delta K_p}^{(i+1,j+1)})$

05. Look up Table I, one can get four 1×7 membership functions matrix $\mu_{L_{\Delta K_p}^{(i,j)}}, \mu_{L_{\Delta K_p}^{(i,j+1)}}, \mu_{L_{\Delta K_p}^{(i+1,j)}}$ and $\mu_{L_{\Delta K_p}^{(i+1,j+1)}}$.

06. Do minimum calculation

$$\tilde{\mu}_{L_{\Delta K_p}^{(i,j)}}^{(n)} = \otimes(\mu_{A_i}, \mu_{B_j}, \mu_{L_{\Delta K_p}^{(i,j)}}^{(n)})$$

$$\tilde{\mu}_{L_{\Delta K_p}^{(i,j+1)}}^{(n)} = \otimes(\mu_{A_i}, \mu_{B_{j+1}}, \mu_{L_{\Delta K_p}^{(i,j+1)}}^{(n)})$$

$$\tilde{\mu}_{L_{\Delta K_p}^{(i+1,j)}}^{(n)} = \otimes(\mu_{A_{i+1}}, \mu_{B_j}, \mu_{L_{\Delta K_p}^{(i+1,j)}}^{(n)})$$

$$\tilde{\mu}_{L_{\Delta K_p}^{(i+1,j+1)}}^{(n)} = \otimes(\mu_{A_{i+1}}, \mu_{B_{j+1}}, \mu_{L_{\Delta K_p}^{(i+1,j+1)}}^{(n)})$$

07. Do maximum calculation

$$\tilde{\mu}_{\Delta K_p}^{(n)} = \oplus(\tilde{\mu}_{L_{\Delta K_p}^{(i,j)}}^{(n)}, \tilde{\mu}_{L_{\Delta K_p}^{(i,j+1)}}^{(n)}, \tilde{\mu}_{L_{\Delta K_p}^{(i+1,j)}}^{(n)}, \tilde{\mu}_{L_{\Delta K_p}^{(i+1,j+1)}}^{(n)})$$

where $n = 1, 2, \ldots, 7$.

08. Defuzzification using the center of gravity method.

$$\Delta K_p = \frac{\sum_{n=1}^{7} \tilde{\mu}_{\Delta K_p}^{(n)} \cdot C_n}{\sum_{n=1}^{7} \tilde{\mu}_{\Delta K_p}^{(n)}}$$

09. Proportional coefficient K_p can be obtained.

$$K_p^{(k)} = K_p^{(k-1)} + K_{\Delta K_p} \cdot \Delta K_p^{(k)}$$

It should be noted K^0_p is the initial proportional coefficient.

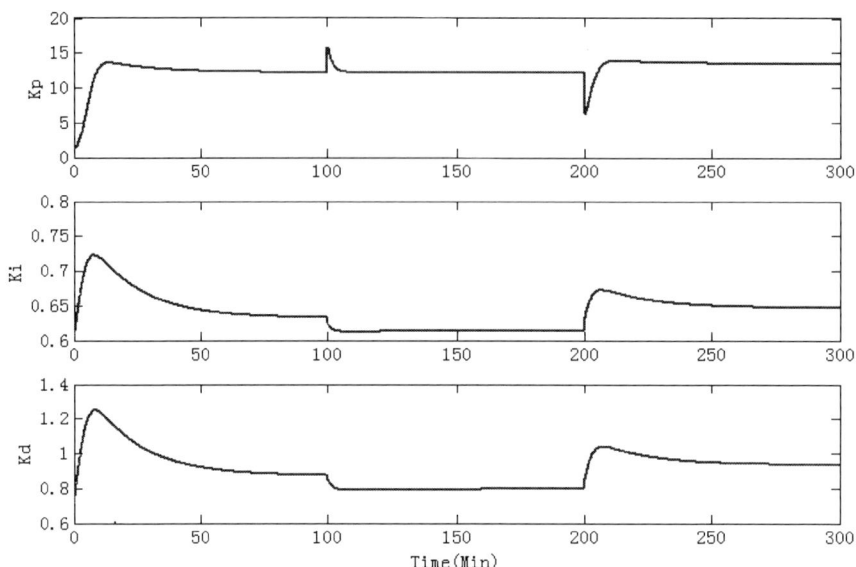

Figure 5. Tuning parameters K_p, K_i, K_d of process.

4. SIMULATIONS AND EXPERIMENTS

To test the fuzzy PID algorithm, the model of the air tank temperature control system is identified with the form

$$T(s) = \frac{0.8127}{26s+1} e^{-0.5s} \tag{2}$$

where the unit of time is minute and the unit of temperature is the centigrade degree (°C).

In the simulations, the temperature set-points of the air tank are changed from 15 °C to 10 °C at the 100th minute, then up to 20°C at the 200th minute (see Figure 4). The parameters of the PID controller are K_p, = 1.4, K_i = 0.6, K_d = 0.7. These parameters are used as the initial K_p, K_i, K_d of the fuzzy PID controller. The scaling gains for error, error change and incremental K_p, are 4e-7, 0.008 and 4.5 respectively. The scaling gains for error, error change and incremental K_i are 0.12, 0.01 and 0.008 respectively. The scaling gains for error, error change and

PID Temperature Control for an Air Tank System ...

incremental K_d are 0.02, 0.01 and 0.02 respectively. From Figure 4, it can be seen that the control performance of the fuzzy PID is better than the PID algorithm. Both dynamic response and steady-state accuracy have been improved. The parameters tuning processes are shown in Figure 5.

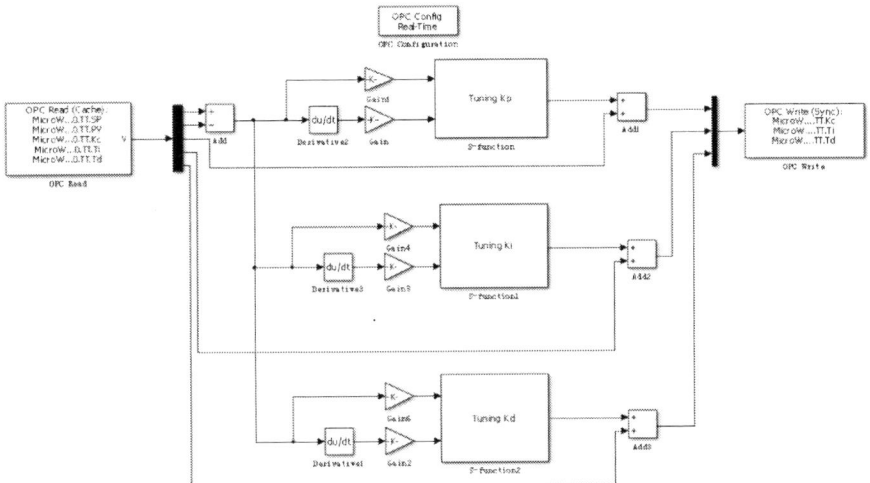

Figure 6. Parameters tuning diagram in Simulink.

In the experiments, the computer loaded Siemens PC Access and Step 7 MicroWin software is located in the room with local control equipment such as Siemens S7-200 PLC, temperature detecting element, frequency transformer and the air tank. The computer loaded Matlab and King View is placed in another room. These two computers are connected with network. IP address of the former is 10.11.5.110 and the latter is 10.11.5.119. Siemens PC Access as an OPC server collects field data from S7-200 and sends signals to the plant through Siemens S7-200. Matlab and KingView as the OPC clients get process data and forward the proportional, integral and differential coefficients to the local controller through the Siemens PC Access. The program of Matlab Simulink is shown in Figure 6, and the supervisory interface of KingView is presented in Figure 7. The real-time experimental results are shown in Figure 8.

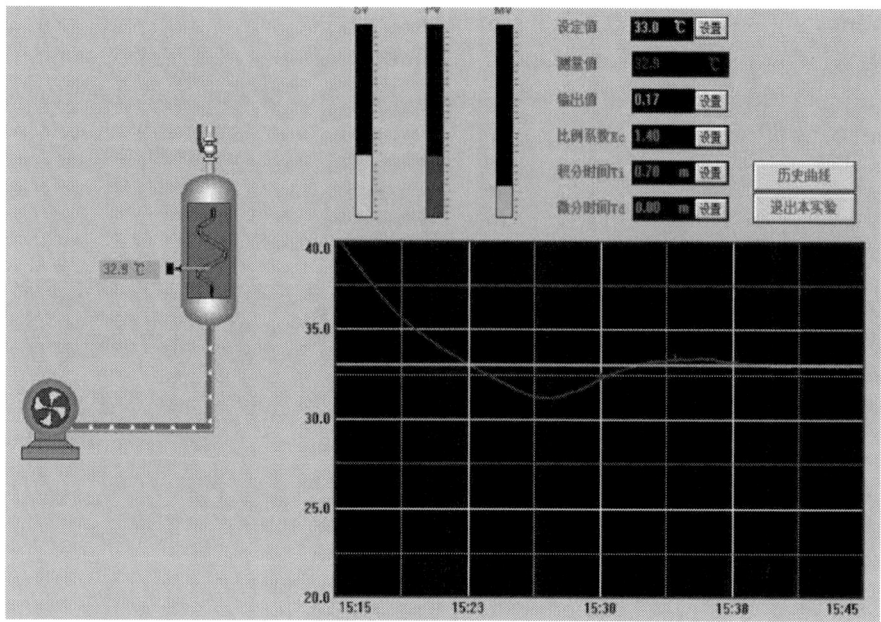

Figure 7. Supervisory interface of King View.

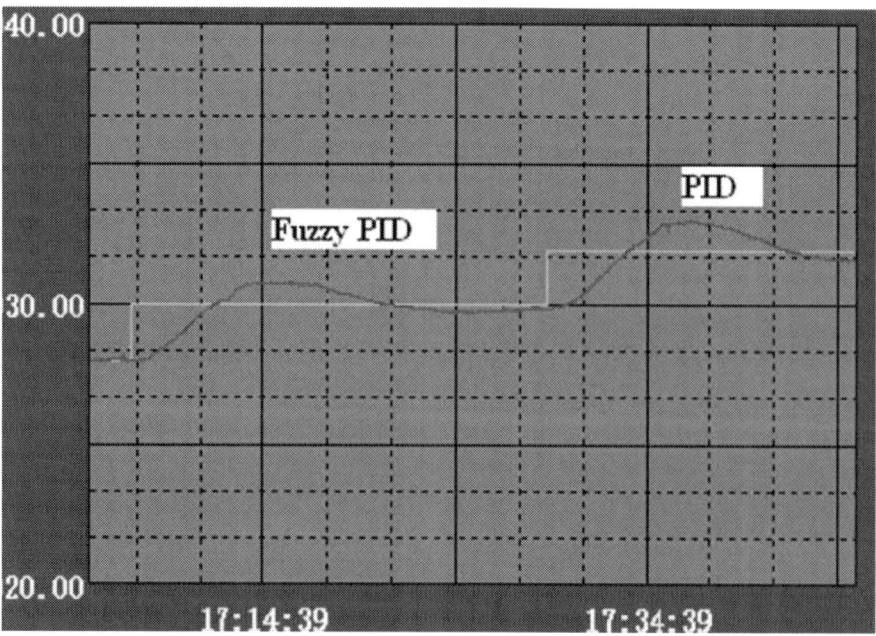

Figure 8. Experimental results comparison between Fuzzy PID and PID.

To reduce the debugging effort, only proportional and integral coefficients are considered in the experiments. The scaling gains for error, error change and incremental K_p are 0.01, 0.01 and 10 respectively and the parameters of PID controller are 1.4, 0.7 for each K_p and K_i. The scaling gains for error, error change and incremental K_i are 0.12, 0.1 and 0.1 respectively. From the comparison between the PID control and the fuzzy PID control of Figure 8, it can be seen that the control performance has been improved with less overshot.

CONCLUSION

This chapter has presented an application of Networked Control in the PID parameters tuning of an air tank temperature control system. Ordinary PID algorithm is utilized in the Siemens PLC to control the temperature of the air tank, and the parameters K_p, K_i, K_d of the PID are regulated by fuzzy inferences which are generated by the Matlab software. The fuzzy controllers are running in the remote computer which is connected to local temperature PLC controller through Network by OPC technique. Both simulations and experiments are given to prove the effectiveness of this control structure. The control performance has been improved much better. Dynamic response time has almost been decreased half time for the simulations. Furthermore, the parameters such as set-points, process values (air tank temperature), K_p, K_i, K_d etc., can be supervised by the configuration software KingView from the remote computers.

REFERENCES

[1] Overstreet J. W., Tzes A. (1999). An internet-based real-time control engineering laboratory, *IEEE Control Systems Magazine*, 19(5), 19-34.

[2] Lazar C., Carari S. (2008). A remote-control engineering laboratory, *IEEE Transactions on Industrial Electronics*, 55(6), 2368-2375.

[3] Hercog D., Gergic B., Uran S., Jezernik K. (2007). A DSP-based remote control laboratory, *IEEE Transactions on Industrial Electronics*, 54(6), 3057-3068.

[4] Casini M., Prattichizzo D., Vicino A. (2003). The automatic control telelab: a user-friendly interface for distance learning, *IEEE Transactions on Education*, 46(2), 252-257.

[5] Garcia-Zubia J., Lopez-de-Ipina D., Orduna P. (2006) Questions and answers for designing useful weblabs, *International Journal of Online Engineering*, 2(3), 1-6.

[6] Hu W. S., Liu G. P., Rees D. (2008). Design and implementation of webbased control laboratory for experiment devices in geographically diverse locations, *IEEE Transactions. on Industrial Electronics*, 55(6), 2343-2354.

[7] Hu W. S., Liu G. P., Rees D. (2008). Design of web-based real-time control laboratory for diversely located test rigs, *Proceedings of the 17th IFAC World Congress*, 12643-12648.

[8] Qiao Y. L., Liu G. P., Zheng G., Hu W. S. (2010). NCSLab: a web-based global-scale control laboratory with rich interactive features, *IEEE Transactions on Industrial Electronics*, 57(10), 3253-3265.

[9] Tong S. W., Qian D. W., Liu G. P. (2014).Networked predictive fuzzy control of systems with forward channel delays based on a linear model predictor, *International Journal of Computers Communications & Control*, 9(4), 471-481.

[10] Tong S. W. (2011). Design and performance analysis of a pH variable domain fuzzy control system, *CAAI Trans. on Intelligent Systems*, 6(4),367-372.

In: Networked Control Systems
Editors: S. Tong and D. Qian

ISBN: 978-1-53619-892-8
© 2021 Nova Science Publishers, Inc.

Chapter 5

ON ZENO BEHAVIOR IN EVENT-TRIGGERED CONTROL OF NETWORKED SYSTEMS

Hao Yu[1,*], PhD, Fei Hao[2] and Tongwen Chen[1]

[1]Department of Electrical and Computer Engineering,
University of Alberta, Edmonton, AB, Canada
[2]School of Automation Science and Electrical Engineering,
Beihang University, Beijing, China

ABSTRACT

In networked systems, Zeno behavior denotes the phenomenon in which an infinite number of transmissions occur in a finite time interval. This phenomenon is extremely undesirable in event-triggered control, which aims at saving communication resources by relating the transmission scheduling with online information. This chapter studies the existence of Zeno behavior in event-triggered control with error-based triggering conditions. It is shown that Zeno behavior is closely related to some particular states, which make the threshold functions in triggering conditions equal to zero. Three kinds of event-triggered control systems, namely, the systems with relative triggering conditions, the finite-time event-triggered control systems, and the systems with

---

[*] Corresponding Author's E-mail: hy10@ualberta.ca.

external threshold signals, are investigated in detail. The corresponding necessary or sufficient conditions for Zeno behavior are obtained. Based on these analyses, it is discovered that an event-triggered control system with a linear plant can be internally stable but not input-to-state stable with respect to external disturbances; some conflicts between finite-time stability and event-triggered control are pointed out; and the difference between the concepts of Zeno-freeness and an event-separation property is revealed. Several numerical examples and simulations are provided to illustrate the feasibility of the proposed results.

Keywords: hybrid systems, event-triggered control, Zeno behavior, finite-time stability

1. INTRODUCTION

Event-triggered control is an aperiodic control scheme that is suitable for networked systems with resource constraints, such as, limited communication bandwidth, restricted energy resources, see, e.g., [1, 2] and the references therein. In event-triggered control, the transmission of information among different nodes is determined by events instead of a fixed period of time. The events are generated based on a triggering condition, which is a criterion involving real-time signals. In this way, event-triggered control can estimate the real-time demand of transmissions, and thus, has potential to provoke transmissions only when necessary. Several pieces of work have demonstrated that event-triggered control can significantly save communication resources without excessive loss of control performance. References [3, 4, 5] studied the asymptotic stability of event-triggered control system from the perspective of input-to-state stability [6]. The event-triggered control with partial state or output feedback was investigated in [7, 8, 9]. In [7, 10, 11], decentralized event-triggered control strategies were studied in distributed networked systems. Several kinds of modeling approaches were proposed in [12, 13] on the event-triggered control for discrete-time plants.

Besides ensuring stability and control performance, one focus in the studies of event-triggered control is to examine the transmission behavior. In [3] and [14], the authors proposed some useful tools to guarantee a positive minimum inter-event time (also known as the event-separation property in [15]). The authors of [15] made a summary on the transmission performance of several popular kinds of event-triggered control systems both in the absence and presence of external disturbances. One reason for the importance in examining transmission behavior of event-triggered control is that event-triggered control systems can be formulated as a hybrid system model (see [16, 17] for details on hybrid systems) as shown in [18, 19, 20, 21]. In fact, if triggering conditions are not designed properly, a particular phenomenon in hybrid systems, called Zeno phenomenon (behavior) [22], might happen. In Zeno phenomenon, an infinite number of discrete transitions (events) take place in a finite time interval. In theory, one solution with Zeno behavior will only have a finite life time; and in practice, a triggering condition that fails to exclude Zeno behavior would require the transmission hardware to eventually work in an infinitely fast frequency. Hence, Zeno behavior is extremely undesirable in event-triggered control.

Several studies have investigated Zeno behavior in hybrid systems. Sufficient conditions [23, 24], and necessary conditions [22] for the existence of Zeno behavior were proposed. Some recent contributions to the study of Zeno behavior can be found in, e.g., [25, 26, 27]. These studies imply that the existence of Zeno behavior are closely related with some particular states, called Zeno equilibrium, the reset map of which directly points to the set of guards. The characterization of Zeno equilibrium was studied in [24] and [27].

Considering the lack of specialized analysis on Zeno behavior in event-triggered control, in this chapter, we will study the conditions on the existence of Zeno behavior and apply them to some well-known event-triggered control schemes. The main contents of this chapter are listed as follows.

First, some necessary and sufficient conditions on the existence of Zeno behavior is provided for a specific form of triggering conditions, which can cover several kinds of existing triggering conditions. Specifically, events are triggered when the difference between the current and the most recently transmitted output signals violates a threshold function that depends on the output signal. Thus, the corresponding Zeno equilibrium set includes all states that lead the threshold function to be zero. By exploiting the property of this Zeno equilibrium set, conditions on the existence of Zeno behavior are given.

Second, as an application of the proposed sufficient conditions, some potential conflicts between finite-time stability and event-triggered control are revealed. It is proved that for unstable plants, the finite-time stability cannot be achieved by the considered triggering condition where the threshold function is independent of time. Our results show that there may exist an enormous obstacle in excluding Zeno phenomenon for finite-time event-triggered control systems. The main challenge is that the finite-time stability is generally achieved by some non-Lipschitz controllers [28]. However, how to ensure simultaneously the local event-separation property and asymptotic stability for the systems with non-Lipschitz controllers is still an open problem. This might explain why there were rare insightful results about this kind of event-triggered control, see, e.g., [29, 30].

Third, by using the proposed necessary condition, the difference of Zeno-freeness and the event-separation property [15] is shown from the study of the triggering conditions involving external threshold signals. In fact, Zeno-freeness is broader than the event-separation property in the sense that it is possible to lose the event-separation property even for a Zeno-free solution. As a result, the transmission performance of a Zeno-free system may also be dissatisfactory. Hence, in some situations, it is not only enough to exclude Zeno phenomenon, but also some tighter conditions are required to guarantee the event-separation property [11].

The remainder of this chapter is organized as follows. Necessary mathematical preliminaries are introduced in Section 2. In Section 3, event-triggered control systems are formulated and the definitions on Zeno behavior are introduced. The main results of this chapter are proposed in Section 4. Then, by using these results, three kinds of event-triggered control systems are investigated in Section 5, namely, systems with the relative triggering condition, finite-time event-triggered control systems, and systems involving external threshold signals. Finally, the conclusions of this chapter are drawn.

2. PRELIMINARIES

In this section, we first introduce some necessary mathematical notations. Then, the concept and relevant definitions on hybrid dynamical systems are provided.

2.1. Notations

Let $\mathbb{R} := (-\infty, \infty)$ and $\mathbb{R}_{\geq 0} := [0, \infty)$. $\mathbb{N}_{\geq 0}$ is defined as the set of the nonnegative integers. The absolute value of a scalar $r \in \mathbb{R}$ is denoted by $|r|$. Euclidian norm of a vector $x \in \mathbb{R}^n$ is denoted by $\|x\|$, and the maximum singular value (Euclidean induced matrixnorm) of a matrix $A \in \mathbb{R}^{n \times m}$ is denoted by $\|A\|$. The transpose of a matrix $A \in \mathbb{R}^{n \times m}$ is denoted by $A^T \in \mathbb{R}^{n \times m}$. For a matrix $A \in \mathbb{R}^{n \times n}$, the maximum and the minimum real parts of its eigenvalues are denoted by $\lambda_{max}(A)$ and $\lambda_{min}(A)$, respectively. A symmetric positive definite matrix $P \in \mathbb{R}^{n \times n}$ is denoted by $P > 0$. The notation sgn($\cdot$) represents the sign function. The space of all essentially bounded functions of dimension n is denoted by L^n_∞. The difference of two sets $\Sigma_1, \Sigma_2 \subset \mathbb{R}^n$ is denoted by $\Sigma_1 \setminus \Sigma_2$, i.e., $\Sigma_1 \setminus \Sigma_2 = \{x \in \mathbb{R}^n \mid x \in \Sigma_1, x \notin \Sigma_2\}$.

A function $f : \mathbb{R}^n \to \mathbb{R}^m$ is Lipschitz continuous on compact sets if for every compactset $S \subset \mathbb{R}^n$ there exists a Lipschitz constant $L > 0$ such that $\|f(x) - f(y)\| \leq L \|x - y\|$ for every $x, y \in S$; otherwise, it is non-Lipschitz (on some compact set). A continuous function $\alpha: \mathbb{R}_{\geq 0} \to \mathbb{R}_{\geq 0}$ belongs to class K_∞ ($\alpha \in K_\infty$) if it is unbounded, zero at zero, and strictly increasing. A function $\alpha : \mathbb{R}^n \to \mathbb{R}_{\geq 0}$ is positive semi-definite if it is nonnegative and satisfies $\alpha(0) = 0$. A function $f : \mathbb{R}^n \to \mathbb{R}^m$ is said to be semi-globally bounded if for any compact set $S \in \mathbb{R}^n$, there exists a constant $M(S) > 0$ such that $\|f(x)\| < M(S)$ for every $x \in S$.

2.2. Hybrid Dynamical Systems

In this chapter, a closed-loop system is formulated as a hybrid system model using the formalism in [16] and [17]. Hence, we first introduce the following hybrid system:

$$\dot{x} = F(x,d), x \in C, x^+ = G(x), x \in D \tag{1}$$

where $x \in \mathsf{R}^n$ is the state vector and $d \in \mathsf{R}^m$ is an external input. The notation x^+ denotes the value of x after one discrete transition. C and D are closed subsets in R^n, and called the flow set and jump set, respectively. F and G are continuous functions from R^n to R^n. To describe the solution of the system in (1), we introduce the following concepts and definitions from [16].

Definition 1. A set $E \subset \mathsf{R}_{\geq 0} \times \mathsf{N}_{\geq 0}$ is called a compact hybrid time domain if $E = \bigcup_{j=0}^{J}([t_j, t_{j+1}], j)$ for some finite sequence of times $0 = t_0 \leq t_1 \leq \cdots \leq t_{J+1}$. Furthermore, E is a hybrid time domain if for arbitrary $(T, J) \in E$, $E \cap ([0, T] \times \{0, 1, \ldots, J\})$ is a compact hybrid time domain.

Definition 2. A hybrid signal is a function defined on a hybrid time domain. Furthermore, a hybrid arc is a hybrid signal $x: \text{dom } x \to \mathsf{R}^n$ such that $x(\cdot, j)$ is locally absolutely continuous for each j.

Definition 3. A hybrid arc $x : \text{dom } x \to \mathsf{R}^n$ is a solution to the hybrid system in (1) if:

1. $\text{dom } x = \text{dom } d$;
2. for $j \in \mathsf{N}_{\geq 0}$ and almost all $t \in \mathsf{R}_{\geq 0}$, which satisfy $(t, j) \in \text{dom } x$, we have $x(t, j) \in C$ and $\dot{x}(t, j) = F(x(t, j), d(t, j))$;
3. for $(t, j) \in \text{dom } x$, which satisfies $(t, j + 1) \in \text{dom } x$, we have $x(t, j) \in D$ and $x(t, j +1) = G(x(t, j))$, where $x(t, j)$ represents the state of the hybrid system after t time units and j jumps.

Based on some suitable assumptions on the data (C, D, F, G), e.g., Proposition S2 in [16], the local existence of solutions from the hybrid system in (1) can be established. For simplicity, we always assume the solution to exist, although it may be non-unique. For an essentially time-continuous state, i.e., $x(t, j+1) = x(t, j)$ for all $(t, j), (t, j+1) \in$ dom x, we sometimes also express it by $x(t)$ with a slight abuse of notation.

3. PROBLEM FORMULATION

Consider the following time-invariant nonlinear plant

$$\dot{x} = f(x, u, d), y = g(x), f(0, 0, 0) = 0, g(0) = 0, \qquad (2)$$

where $x \in R^n$ is the state vector and $u \in R^m$ is a control input. $y \in R^q$ is the measurable output. $d \in L^p_\infty$ denotes the bounded external disturbance with a bound $d_{max} \geq 0$. The functions $f: R^n \times R^m \times R^p \to R^n$, $g: R^n \to R^p$ are C^1 continuous; and thus, they are semi-globally bounded and Lipschitz continuous on compact sets. Especially, $\partial g(x)/\partial x$ is semi-globally bounded. Without loss of generality, we suppose the initial instant as $t_0 = 0$.

Due to limited communication resources in networked systems, the controller is implemented in an event-triggered manner, i.e.,

$$u = \kappa(y(t_k)), \text{ for } t \in [t_k, t_{k+1}) \qquad (3)$$

where $\kappa: R^q \to R^m$ is a semi-globally bounded function with $\kappa(0) = 0$. Especially, κ is not required to be Lipschitz continuous on compact sets. The monotonically increas-ing sequence $\{t_k\}^\infty_{k=0}$ represents the triggering times decided by the following triggering condition:

$$t_{k+1} = \inf\{t \geq t_k \mid \|e_y(t)\| > g_0(y(t))\} \qquad (4)$$

where $e_y(t) := y(t_k) - y(t)$, $t \in [t_k, t_{k+1})$, is the measurement error. The threshold function $g_0: R^q \to R_{\geq 0}$ is continuous and assumed to have the following two properties.

Assumption 1. *The triggering function* g_0 *in* (4) *satisfies*

1. *norm-invariance*: $g_0(y_1) = g_0(y_2)$ for any $y_1, y_2 \in R^q$ satisfying $\|y_1\| = \|y_2\|$;
2. *triangle inequality*: $g_0(y_1 + y_2) \leq g_0(y_1) + g_0(y_2)$ holds for any $y_1, y_2 \in R^q$.

In addition, we introduce the concept of local compression on the threshold function g_0.

Definition 4 *The threshold function $g_0(\cdot)$ is locally compressed if there exists no positive constant $\varepsilon_y > 0$ such that $g_0(y) \geq \|y\|$ for all $y \in \{s \in R^q \mid 0 < \|s\| \leq \varepsilon_y\}$.*

The triggering condition in (4) satisfies that, for any two different triggering instants t_{s1} and t_{s2}, if the corresponding triggering states satisfy $x(t_{s1}) = x(t_{s2}) =: x_m$, then they lead to the same set $\Lambda(x_m)$ that includes all the states that cannot trigger events, i.e.,

$$\Lambda(x_m) := \{x \in R^n \mid \|g(x) - g(x_m)\| \leq g_0(g(x))\}.$$

Thus, the triggering condition in (4) is time-invariant since $\Lambda(\cdot)$ is independent of time. Meanwhile, this property ensures that, for any triggering instant t_k, the set $\Lambda(x(t_k))$ relies on only the state $x(t_k)$ rather than the state trajectories within the inter-event interval (t_k, t_{k+1}). Thus, the triggering condition in (4) belongs to the category of static event-triggered schemes.

Remark 1. The considered triggering conditions can cover several kinds of triggering conditions in existing studies. For example, when $g_0(y)$ is a positive (semi-)definite function, (4) denotes the relative triggering condition employed in [3, 8]. When $g_0(y)$ is equal to a positive constant, (4) denotes the absolute triggering condition introduced in [14]. For example, the following two triggering conditions:

$$t_{k+1} = \inf\{t \geq t_k \mid \|e_y(t)\| > 1/2 \, \|y(t)\|\}, \text{ and, } t_{k+1} = \inf\{t \geq t_k \mid \|e_y(t)\| > 1/2\}$$

are relative and absolute, respectively.

Then, we formulate the closed-loop event-triggered control system as a hybrid system model,

$$\begin{cases} \dot{x} = f(x, \kappa(y+e_y), d), & [x^T, e_y^T]^T \in C, \\ \dot{e}_y = -\dfrac{\partial g(x)}{\partial x} \cdot f(x, \kappa(y+e_y), d), & [x^T, e_y^T]^T \in C, \\ x^+ = x, & [x^T, e_y^T]^T \in D, \\ e_y^+ = 0, & [x^T, e_y^T]^T \in D, \\ y = g(x), \end{cases} \qquad (5)$$

where the flow set is $C := \{[x^T, e^T_y]^T \in \mathbb{R}^{n+q} \mid \|e_y\| \leq g_0(g_2(x))\}$, and the jump set is $D := \{[x^T, e^T_y]^T \in \mathbb{R}^{n+q} \mid \|e_y\| \geq g_0(g_2(x))\}$. In addition, we assume that there exists a mandatory jump at the initial instant 0, which yields $e_y(0, 0) = 0$. Note that since κ is not necessarily Lipschitz continuous, $f(x, \kappa(g(x) + e_y), d)$ may be not Lipschitz on compact sets with respect to x, e_y, d, either.

To characterize the properties of triggering time sequences, we initially introduce the mathematical definitions on Zeno behavior.

Definition 5. ([23]) *For a given initial state $x(0) = x_0$ and a bounded disturbance signal $d(t)$, define $\chi(x_0, d(t))$ as a solution of the event-triggered control system in (5). Then, a solution χ is Zeno if there exists a nonnegative constant $t_\infty < \infty$ such that*

$$\lim_{k \to \infty} t_k = \sum_{k=0}^{\infty} (t_{k+1} - t_k) = t_\infty$$

where t_∞ is called a Zeno instant. Otherwise, the solution is Zeno-free. An event-triggered control system is Zeno if it possesses a Zeno solution for some initial state and disturbance signal.

Definition 5 leads to two different kinds of Zeno behavior [23], which are defined as follows.

Definition 6. *For a Zeno solution χ, it is*
1. *Chattering Zeno: if there exists a finite integer $L > 0$ such that $t_{k+1} - t_k = 0$ for all $k \geq L$.*
2. *Genuinely Zeno: if $t_{k+1} - t_k > 0$ for all $k \geq 0$.*

Remark 2. The difference between these two kinds of Zeno behavior can be reflected in their analysis and detection [23]. Simply speaking, chattering Zeno solutions result from only the jump of systems while genuinely Zeno solutions are generated by the flow and jump behavior. Thus, chattering Zeno behavior is easier to be detected than genuinely Zeno one. Moreover, it is obvious that a solution cannot be chattering Zeno and genuinely Zeno simultaneously. Then, we will analyze these two kinds of Zeno behavior separately.

Several studies, e.g., [22, 27, 31], implied that Zeno behavior is closely related to a class of special state sets, which are called Zeno equilibria.

Definition 7. *A Zeno equilibrium set of the event-triggered control system in (5) with the triggering condition in (4) is*

$$\Gamma := \{x \in R^n \mid g_0(g(x)) = 0\}. \tag{6}$$

From Definition 7, the Zeno equilibrium set consists of all the states that set the threshold function to zero. Similarly, the Zeno equilibrium set in (6) is time-invariant since Γ is independent of the current instant t. Furthermore, based on the form of triggering conditions in (4), one can directly obtain the following property of Zeno equilibria.

Lemma 1. *Consider the triggering condition in (4) and two triggering instants t_{s1} and t_{s2} that satisfy $g(x(t_{s1})) = g(x(t_{s2}))$. If $x(t_{s1})$ belongs to the Zeno equilibrium set Γ, then $x(t_{s2}) \in \Gamma$ as well.*

For further analysis of Zeno behavior, we introduce the following two assumptions on the closed-loop system in (5).

First, for a stable plant, the trivial controller $u = 0$ is enough to stabilize the plant and no transmission is required. Thus, sometimes we only consider the unstable plant. Specifically, the plant in (2) should not be bounded-input-bounded-state stable in the following sense.

Assumption 2. *Consider a given controller function κ and disturbance bound $d_{max} \geq 0$. For any constant $\delta > 0$, there always exist $x_0 \in \{x \in R^n \mid \|x\| \leq \delta\}$ and $d(t)$, satisfying $\|d(t)\| \leq d_{max}$, $t \in R_{\geq 0}$, such that the solution of $\dot{x}(t) = f(x(t), \kappa(g(x_0)), d(t))$, under the initial state $x(0) = x_0$, is unbounded. Define the set of all such x_0 as $\Gamma_x(\kappa, d_{max})$.*

Moreover, to stabilize an unstable plant, only some properly designed controllers and triggering conditions should be considered. A fundamental requirement is that the controller can keep the state x from divergence if the triggering conditions are not violated. This is summarized in the following assumption.

Assumption 3. *The controller in (3) and triggering condition in (4) are supposed to be well-designed, i.e., for all $d(t) \in L^p_\infty$, they can lead the corresponding hybrid system in (5) to admit a C^1 function $V: \mathbb{R}^n \rightarrow \mathbb{R}_{\geq 0}$ such that there exists a positive constant $M > 0$, and $\phi_1, \phi_2 \in K_\infty$ satisfying*

1. $\phi_1(\|x\|) \leq V(x) \leq \phi_2(\|x\|)$;
2. $V(x) \geq M \Rightarrow \dot{V} := \partial V / \partial x \cdot f(x, \kappa((y + e_y)), d) \leq 0$,

for all $[x^T, e^T_y]^T \in C$.

At first glance, Assumption 3 provides a candidate Lyapunov function of (5) and shows that the closed-loop system in (5) is stable. However, the assumption only involves the flow of the x-subsystem. Thus, Assumption 3 merely implies the boundedness since V does not increase in both flow and jump sets when $V(x) > M$. Neither the ultimate boundedness nor the asymptotic stability can be obtained from the assumption since it tells nothing about the transmission behavior.

Therefore, the main interest of this chapter is to propose necessary and sufficient conditions on the existence of Zeno behavior for the considered event-triggered control system in (5). Using these conditions, the transmission behavior of several different event-triggered control systems is examined and the corresponding results on Zeno behavior are proposed.

4. MAIN RESULTS

In this section, we propose the main results of this chapter. As in the analysis in Remark 2, chattering Zeno and genuinely Zeno behavior will be studied separately. According to the triggering conditions in (4), there must be no event in the interval that the output $y(t)$ can keep constant under the control input $\kappa(g(x(t_k)))$. Thus, before giving the conditions on chattering Zeno behavior the following definition of a local output equilibrium is introduced.

Definition 8. Consider the system in (5) under a disturbance $d(t) \in L^{p_\infty}$. The local output equilibrium set at time $h \in R_{\geq 0}$ is define as $\Gamma_{d(t)}(h) := \{\bar{x} \in R^n \mid \text{there exists a constant } \epsilon_y > 0 \text{ such that } g(x(t)) = g(\bar{x}), t \in [0, \epsilon_y)$, where $x(t)$ is the solution of the differential equation $\dot{x}(t) = f(x(t), \kappa(g(\bar{x})), d(t+h))$ under the initial state $x(0) = \bar{x}\}$

Based on the definitions above, the following theorem provides results on chattering Zeno behavior.

Theorem 1. Under Assumption 1, a solution χ of the event-triggered control system in (5) with a disturbance $d(t)$ is chattering Zeno if and only if there exists a triggering instant $t_{k0} \geq 0$ such that $x(t_{k0}, k_0) \in \Gamma \setminus \Gamma_{d(t)}(t_{k0})$, and g_0 is locally compressed.

Proof. We first show sufficiency. Since $x(t_{k0}, k_0) \in \Gamma$, one has $g_0(g(x(t_{k0}, k_0))) = 0$. Now suppose that $x(t_{k0}, k_0)$ will not trigger another event immediately, i.e., $t_{k0+1} \neq t_{k0}$. As a result, there must exist a positive constant ϵ_1 such that $\|e_y(t, k_0)\| \leq g_0(y(t, k_0))$ for $t \in [t_{k0}, t_{k0} + \epsilon_1]$. Furthermore, we have

$$\begin{aligned}
\|e_y(t,k_0)\| &= \|y(t,k_0) - y(t_{k_0},k_0)\| \\
&\leq g_0(y(t,k_0)) \\
&\stackrel{(i)}{\leq} g_0(y(t,k_0) - y(t_{k_0},k_0)) + g_0(y(t_{k_0},k_0)) \\
&\stackrel{(ii)}{=} g_0(e_y(t,k_0))
\end{aligned} \quad (7)$$

for $t \in [t_0, t_0 + \epsilon_1]$, where (i) utilizes the triangle inequality of g_0 and (ii) is due to the facts that $x(t_{k0}, k_0) \in \Gamma$ and the Zeno equilibrium set is time-invariant. Since $x(t_{k0}, k_0) \notin \Gamma_{d(t)}(t_{k0})$, there exists no constant $\epsilon_2 > 0$ such that $e_y(t, k_0) = 0$ for all $t \in (t_{k0}, t_{k0} + \epsilon_2]$. Due to the continuity of the states on the flow, there must exist positive constants $0 < \epsilon_3 \leq \epsilon_1$ and $\epsilon_y > 0$ such that $e_y(t, t_0) \in \{z \in R^q \mid 0 < \|z\| \leq \epsilon_y\}$, $t \in (t_0, t_0 + \epsilon_3)$. As a result, (7) contradicts the local compression property of g_0 due to the norm invariance, and then, t_{k0+1} must be equal to t_{k0}.

As $t_{k0+1} = t_{k0}$, $x(t_{k0}, k_0 + 1) = x(t_{k0+1}, k_0 + 1)$ must belong to the solution of the event-triggered control system in (5). Since the x-subsystem does not jump, $x(t_{k0}, k_0 + 1)$ also belongs to $\Gamma \setminus \Gamma_{d(t)}(t_{k0})$.

Repeating the above process, one has that for any $i \geq 0$, $x(t_{k0}, k_0 + i)$ belongs to the solution; i.e., $\lim_{x \to \infty} t_{k_0+i} = t_{k_0}$ and $t_{j+1} - t_j = t_{k0} - t_{k0} = 0$ for all $j \geq k_0$. Therefore, this solution is chattering Zeno with a Zeno instant t_{k0} and the sufficiency is proved.

Next, we consider necessity. Since the solution is chattering Zeno, there exists a Zeno instant $t_\infty \geq 0$ and a constant $L \geq 0$ such that $\lim_{i \to \infty} t_{L+i} = t_\infty$ and $t_{k+1} - t_k = 0$ for all $k \geq L$, which yield that $t_{L+i} = t_\infty$ for all $i \geq 0$. Thus,

$$g_0(g(x(t_{L+i+1}, L+i))) = g_0(g(x(t_\infty, L+i+1)))$$
$$= \| g(x(t_{L+i}, L+i)) - g(x(t_{L+i+1}, L+i)) \|$$
$$= \| g(x(t_\infty, L+i)) - g(x(t_\infty, L+i)) \|$$
$$= 0.$$

for all $i \in \mathbb{N}_{\geq 0}$. This implies that $x(t_{L+i}, L + i + 1) \in \Gamma$ for all $i \geq 0$. Since the x-subsystem does not jump essentially, $x(t_\infty, L) = x(t_\infty, L + i + 1) \in \Gamma$, for all $i \geq 0$. Let $t_{k0} = t_\infty$ and $k_0 = L + 1$. If $x(t_{k0}, k_0) \in \Gamma \cap \Gamma_{d(t)}(t_{k0})$, there must exist a constant $\varepsilon_4 > 0$ such that $0 = \|e_y(t, k_0)\| \leq g_0(g(x(t, k_0)))$ for $t \in (t_{k0}, t_{k0} + \varepsilon_4]$ which contradicts that t_0 is a chattering Zeno instant

Moreover, if g_0 is not locally compressed, there exists a positive constant ε_y such that $\|y\| \leq g_0(y)$ for all $y \in \{y \in \mathbb{R}^q \mid 0 < \|y\| \leq \varepsilon_y\}$. Then, one has

$$\begin{aligned} g(y(t,k_0)) &\stackrel{(i)}{\geq} g_0(y(t,k_0) - y(t_{k0},k_0)) - g(-y(t_{k0},k_0)) \\ &\stackrel{(ii)}{=} g_0(e_y(t,k_0)) - g_0(y(t_{k0},k_0)) \\ &\stackrel{(iii)}{=} g_0(e_y(t,k_0)) \\ &\geq \| e_y(t,k_0) \| \end{aligned} \quad (8)$$

for $e_y \in \{z \in \mathbb{R}^q \mid 0 < \|z\| \leq \varepsilon_y\}$. (i) and (ii) can be obtained by, respectively, the triangle inequality and the norm-invariance of g_0; while (iii) is due to the facts that $x(t_{k0}, k_0) \in \Gamma$ and the Zeno equilibrium set is time-invariant. (8) implies that the event would not be triggered if $e_y \in \{z \in \mathbb{R}^q \mid 0 \leq \|z\| < \varepsilon_y\}$. Since $x \notin \Gamma_{d(t)}(t_{k0})$ and the system is continuous on the flow, there must exist a positive constant $\varepsilon_5 > 0$ such that $e_y(t, k_0) \in \{z \in \mathbb{R}^q \mid 0 < \|z\| < \varepsilon_y\}$, for all $t \in (t_{k0}, t_{k0} + \varepsilon_5)$. Therefore, $t_{k0+1} \neq t_{k0}$, which contradicts that t_{k0} is a chattering Zeno instant. Thus, if the solution is chattering Zeno, g_0 is necessarily locally compressed and the proof is completed.

Then, the conditions of genuinely Zeno behavior are characterized in the following theorem.

Theorem 2. Under Assumptions 1 and 3 and the conditions of Lemma 1, a solution x of the event-triggered control system in (5) is genuinely Zeno if and only if there exist a triggering instant $t_{k0} \in \mathbb{R}_{\geq 0}$ and a constant $T_0 > t_{k0}$ such that $x(t_{k0}, k_0) \notin \Gamma$ and $\lim_{t \to T_0} x(t) = x_\Gamma$ with some $x_\Gamma \in \Gamma$.

Proof. For sufficiency, without loss of generality, we assume that $t_{k0} = k_0 = 0$ and T_0 is the first time that the state of x-subsystem arrives at the Zeno equilibrium set, i.e., $x(t) \notin \Gamma$ for all $t \in [t_0, T_0)$ and $\lim_{t \to T_0} x(t) = x_\Gamma$. We prove the conclusion by contradiction. Assume that this solution is not genuinely Zeno. Since $x(t) \notin \Gamma$ for $t \in [t_0, T_0)$, from Theorem 1, the solution in the interval $[t_0, T_0)$ does not exhibit chattering Zeno behavior, either. Subsequently, there must exist finite triggering instants during the interval $[t_0, T_0)$. Define $K_0 = \text{argmax}_k \{t \in \{t_k\}_{k=0}^\infty \mid t < T_0\}$, then there exists a positive constant $\delta > 0$ such that $T_0 - t_{K0} = \delta$. From the triggering condition in (4), it follows that, for any $t \in [t_{K0}, T_0)$, $\|y(t_{K0}) - y(t)\| \leq g_0(g(x(t)))$. Due to the continuity of x, g and g_0, one has

$$\| g(x((t_{K_0}))) - g(x_\Gamma) \| = \left\| y(t_{K_0}) - \lim_{t \to T_0} y(t) \right\|$$

$$= \lim_{t \to T_0} \| y(t_{K_0}) - y(t) \|$$

$$\leq \lim_{t \to T_0} g_0(g(x(t)))$$

$$= g_0(g(x_\Gamma)) = 0.$$

This implies that $g(x(t_{K0})) = g(x_\Gamma)$. Since $x_\Gamma \in \Gamma$, from Lemma 1, it follows $x(t_{K0}) \in \Gamma$. This contradicts the fact that $x(t) \notin \Gamma$ for all $t \in [t_0, T_0)$.

For necessity, we first prove that if there exists a triggering instant t_{k0} such that $x(t_{k0}, k_0) \notin \Gamma$, then a genuinely Zeno solution necessarily arrives at the Zeno equilibrium set in a finite time interval. Without loss of generality, we assume $t_{k0} = 0$ and $x(0, 0) \notin \Gamma$. Since the solution is genuinely Zeno, there exists a Zeno instant $t_\infty > 0$ such that $\lim_{k \to \infty} t_k = t_\infty$. From the triggering condition in (4), it follows that, for any consecutive triggering instants t_k and t_{k+1},

$$\| y(t_k) - y(t_{k+1}) \| = \| e_y(t_{k+1}, k) \| = g_0(g(x(t_{k+1}))).$$

Subsequently, we studied the following two sequences:

$T_1 := \{x(t_1), x(t_2), \ldots, x(t_k). \ldots \}$,

$T_2 := \{\| e_y(t_1, 0) \|, \| e_y(t_2, 1) \|, \ldots, \| e_y(t_k, k-1) \|, \ldots \}$.

Because of Assumption 3, for any solution with $x(0) \notin \Gamma$, there exists a compact set $S \subset R^n$ such that $x(t) \in S$ for all $t \geq 0$. Due to the fact that f, g and κ are semi-globally bounded, there must exist a positive constant $M_1 > 0$ such that $\| f(x(t), \kappa(y(t_k)), d(t)) \| < M_1$ for all $t \in (t_k, t_{k+1})$ and $k \in N_{\geq 0}$. Similarly, since $\frac{\partial g(x)}{\partial x}$ is semi-globally bounded, there also exists a positive constant $M_2 > 0$ such that $\| \dot{e}_y(t) \| < M_2$ for all $t \in (t_k, t_{k+1})$ and $k \in \mathbb{N}_{\geq 0}$, where $\dot{y}(t) = \dfrac{d}{dt} g(x(t)) = \dfrac{\partial g(x)}{\partial x} \cdot f(x(t), \kappa(y(t_k)), d(t))$.

Since $\lim_{k \to \infty} t_k = t_\infty$, for any given $\varepsilon > 0$, there must exist a positive integer $H(\varepsilon) \in \mathbb{N}_{\geq 0}$ such that, for any integers $k_1, k_2 > H(\varepsilon)$, one has $|t_{k_1} - t_{k_2}| < \varepsilon$. Furthermore, define $H_x := H(\dfrac{\varepsilon_x}{M_1})$. Then, for any integers $m > n \geq H_x$,

$$\| x(t_m) - x(t_n) \| \leq (t_m - t_n) M_1 \leq \dfrac{\varepsilon_x}{M_1} M_1 = \varepsilon_x \tag{9}$$

According to the Cauchy convergence criterion, (9) implies that $x(t_k)$ converges as $k \to \infty$, i.e., there exists $x_\Gamma \in \mathbb{R}^n$ such that $\lim_{k \to \infty} x(t_k) = x_\Gamma$.

Similarly, for any $\varepsilon_e > 0$, let $H_e := H(\dfrac{\varepsilon_e}{M_2})$. Then, for any integer $k \geq H_e$

$$\| e_y(t_{k+1}, k) \| = \left\| \int_{t_k}^{t_{k+1}} \dot{y}(\tau, k) d\tau \right\| \leq (t_{k+1} - t_k) M_2 \leq \dfrac{\varepsilon_e}{M_2} M_2 = \varepsilon_e,$$

which implies that $\lim_{k \to \infty} \|e_y(t_k, k - 1)\| = 0$.

Then we consider the following sequence:

$$T_3 := \{g_0(g(x(t_1))), g_0(g(x(t_2))), \ldots, g_0(g(x(t_k))), \ldots\}.$$

The triggering condition in (4) ensures $T_2 = T_3$. Hence, $\lim_{k \to \infty} g_0(g(x(t_k))) = 0$. Due to the continuity of g_0, g and x, one has that

$$\lim_{k \to \infty} g_0(g(x(t_k))) = g_0(g(\lim_{k \to \infty} x(t_k))) = g_0(g(x_\Gamma)) = 0$$

which yields $x_\Gamma \in \Gamma$. Thus, we have proved that the states at the triggering instants will arrive at the Zeno equilibrium set eventually.

Next, we study the state trajectories of the x-subsystem during triggering instants. For any $t \in [t_k, t_{k+1})$,

$$\|x(t) - x_\Gamma\| = \left\|\int_{t_k}^t \dot{x}(\tau, k)d\tau + x(t_k) - x_\Gamma\right\| \leq M_1(t_{k+1} - t_k) + \|x(t_k) - x_\Gamma\|$$

Since $\lim_{k\to\infty}(t_{k+1} - t_k) = 0$ and $\lim_{k\to\infty}\|x(t_k) - x_\Gamma\| = 0$, one has that $\lim_{t\to\infty}\|x(t_k) - x_\Gamma\| = 0$. Thus, in the theorem, one can select $T_0 = t_\infty$.

Then, we prove by contradiction that at least one triggering state of a genuinely Zeno solution is necessarily outside the Zeno equilibrium set. Suppose this condition does not hold, hence $x(t_k, k) \in \Gamma$ for all $k \in \mathbb{N}_{\geq 0}$.

Step 1: proving $x(t_k, k) \in \Gamma \cap \Gamma_{d(t)}$ (t_k), $k \in \mathbb{N}_{\geq 0}$.

If $x(0, 0) \in \Gamma\Gamma_{d(t)}$ (0), then we have that g_0 cannot be locally compressed. Otherwise, Theorem 1 ensures this solution to be chattering Zeno, which contradicts the fact that this solution is genuinely Zeno. Consequently, there exists a constant $\tau_s \in (0, t_1)$ such that $x(0, \tau_s) \notin \Gamma$ and $\lim_{t\to t_1} x(t, 0) \in \Gamma$. Considering that g_0 is not locally compressed, we have that there exists a scalar $\varepsilon_y > 0$ such that $g_0(e_y) \geq \|e_y\|$ for all $e_y \in \{z \in \mathbb{R}^q \mid 0 \leq \|z\| \leq \varepsilon_y\}$. From the triggering condition in (4) and the fact of $x(t_1, 0) \in \Gamma$, it follows that $y(t_0, 0) = y(t_1, 0)$. Hence, by using a similar analysis in (8), one can show that the event cannot be triggered at instant t_1, which contradicts the fact that t_1 is a triggering instant. Repeating the process above, one can prove $x(t_k, k) \in \Gamma \cap \Gamma_{d(t)}$ (t_k), $k \in \mathbb{N}_{\geq 0}$. Meanwhile, the analysis also implies that it is necessary for g_0 to be locally compressed if the considered $\tau_s \in (t_k, t_{k+1})$ exists for some $k \in \mathbb{N}_{\geq 0}$.

Step 2: proving $y(t) = y(0)$ for any $t \in [0, t_2)$.

From the definition of $\Gamma_{d(t)}$ (h), it follows that there exist constants ϵ_k, $k = 1, 2$, satfying $y(t, k) = y(t_k, k)$ for $t \in [t_k, \epsilon_k]$. Without loss of generality, assume ϵ_k as the largest value in $(t_k, t_{k+1}]$ to satisfy the relationship during the k-th inter-event interval with $k = 1, 2$. It is obvious that $\epsilon_0 < \epsilon_1$ Then, we prove that $\epsilon_k = t_{k+1}$ for $k = 0, 1$. If there exists no $\tau_{\epsilon k} \in (t_k, t_{k+1})$ such that $e_y(\tau_{\epsilon k}, k) \neq 0$, then $\epsilon_k = t_{k+1}$. If such $\tau_{\epsilon k}$

exists, it can be proved that g_0 must not be locally compressed. Otherwise, the analysis in (8) implies that $e_y(\tau_{\epsilon k}, k) \neq 0$ will lead to an extra triggering instant, which contradicts the fact of $\tau_{\epsilon k} \in (t_k, t_{k+1})$ and proves that g_0 is not locally compressed. Meanwhile, the definition of $\tau_{\epsilon k}$ yields $x(\tau_{\epsilon k}) \notin \Gamma$; thus, $\tau_{\epsilon k}$ has the same meaning as τ_s in Step 1. From the analysis in Step 1, g_0 should be locally compressed due to the fact of $x(t_k) \in \Gamma$, $k \in \mathbb{N}_{\geq 0}$. Since there is no function that can be locally compressed and not be locally compressed simultaneously, a contradiction is achieved. Thus, based on the definition of ϵ_k, one has $y(t) = y(0)$ for any $t \in [0, t_2)$.

Step 3: proving that t_1 cannot be a triggering instant.

According to Step 2, there is no change in the output and input just before and after the instant t_1 since $y(t) = y(0)$ for any $t \in [0, t_2)$. Meanwhile, note that there is no event during $t \in (t_1, t_2)$. Thus, even if there is no transmission at the instant t_1, the event will also not be triggered any more. Thus, t_1 should not be a triggering instant.

Repeating the last two processes above, one can show that after the initial event at $t = 0$, there will be no further events, which contradicts that the solution is genuinely Zeno. Therefore, the necessity is proved.

Remark 3. Compared to Theorem 1, Theorem 2 does not require the hypotheses of local compression of g_0 and $x_\Gamma \notin \Gamma_{d(t)}(t_\infty)$. Since a Zeno solution, with the initial state outside the Zeno equilibrium set, would never really arrive at the Zeno points, the properties on the Zeno points are not necessarily considered. Analogously, since chattering Zeno behavior is caused by the jump, the property on the flow, such as Assumption 3, is not required in Theorem 1.

Remark 4. To check the conditions of Theorems 1 and 2, one needs to calculate the Zeno equilibrium set Γ and local output equilibrium set $\Gamma_{d(t)}(h)$. The calculation of Γ requires to solve the (nonlinear) equation $g_0(g(x)) = 0$. Since the threshold function is given by designers, in the case that the output function g is not complicated (for example, g is a linear combination of states), the designer can use a more conservative but simpler threshold function g_0 to simplify the calculation of Γ. On the other hand, it is usually difficult to compute the local output equilibrium set since it involves unknown disturbances. However, the introduction of $\Gamma_{d(t)}(h)$ is to "fill in the gap" between the sufficient and necessary conditions. Thus, if designers focus only on the unilateral condition, the calculation of $\Gamma_{d(t)}$ (h) is not necessary. For example, a necessary condition for Zeno behavior is $\Gamma \neq \emptyset$, which does not depend on $\Gamma_{d(t)}$ (h). In addition, although the proof of Theorem 2 involves the local output equilibrium set, its statement is independent of $\Gamma_{d(t)}$ (h). Meanwhile, there generally exists no state that belongs to the local output equilibrium set for any considered bounded disturbances. Thus, the conditions of Theorem 1 on the local output equilibrium set can always be assumed to hold under some particular disturbance.

Remark 5. In some studies, the triggering conditions were defined by non-strict inequalities, such as the ones considered in [32]:

$$t_{k+1} = \inf\{t \geq t_k | \; ||e_y(t)|| \geq g_0(y(t))\}, \tag{10}$$

where the sign ">" in (4) is replaced by "≥". Although both (4) and (10) lead to the same form of hybrid system in (5), they have different effects on the transmission performance. Note that this is possible since the solution of a hybrid system may be non-unique as introduced in Section 2. First, for (10), an event will be always triggered whenever the state $[x^T, e^T_y]^T$ touches the boundary of the flow set C; while for (4), only the state that really tends to go across from the flow set C to the jump set D can lead to events. Second, by using similar analysis as in Theorems 1 and 2, one can obtain that the sufficient and necessary condition on Zeno behavior for (10) is independent of the local output equilibrium set. Especially, all states in the Zeno equilibrium set would lead to chattering Zeno solutions regardless of the dynamics of the system. Thus, this chapter considers a more general and effective form of triggering conditions.

It is worth noting that the property in Lemma 1 plays a key role in the proof of the sufficiency in Theorem 2, which is illustrated by the following example.

Example 1. Consider the following system:

$$\begin{cases} \dot{x}_a = -\text{sgn}(x_a)\sqrt{|x_a|} + u \\ \dot{x}_b = -x_b, \end{cases} \qquad y = x_b, u = -x_b(tk)$$

Define the triggering condition as

$$t_{k+1} = \inf\{t \geq t_k \mid |e_y(t)| > \delta\sqrt{|x_a(t)|}\}$$

where $\delta \in (0, 1)$. The corresponding Zeno equilibrium set is $\{x = [x_a, x_b]^T \in \mathbb{R}^2 \mid x_a = 0\}$, which does not meet the condition in Lemma 1 obviously.

Choosing the function in Assumption 3 as

$$V = \frac{2}{3}|x_a|^{\frac{3}{2}} + \frac{c}{2}x_b^2$$

where $c > 0$, we have

$$\begin{aligned}
\dot{V} &= \operatorname{sgn}(x_a)\dot{x}_a\sqrt{|x_a|} + c\dot{x}_b x_b \\
&= -\operatorname{sgn}^2(x_a)|x_a| - \operatorname{sgn}(x_a)\sqrt{|x_a|}x_b - \operatorname{sgn}(x_a)\sqrt{|x_a|}e_y - cx_b^2 \\
&\leq -|x_a| + \sqrt{|x_a|}|x_b| - cx_b^2 + \sqrt{|x_a|}|e_y| \\
&\leq -(1-\delta)|x_a| + \sqrt{|x_a|}|x_b| - cx_b^2 \\
&\leq -(1-\delta)|x_a| + (\frac{1-\delta}{2})|x_a| + \frac{1}{2-2\delta}x_b^2 - cx_b^2 \\
&\leq -\frac{1-\delta}{2}|x_a| - (c - \frac{1}{2-2\delta})x_b^2
\end{aligned}$$

Hence, for any given $\delta \in (0, 1)$, one can find $c > 0$ such that $\dot{V} \leq 0$, therefore, Assumption 3 holds with $M = 0$.

The initial state is supposed to be $x_a(0) = 1$ and $x_b(0) = 0$ which are not in the Zeno equilibrium set. The corresponding solution is

$$\begin{cases} x_a(t) = (1 - \frac{t}{2})^2, t \in [0, 2), \\ x_a(t) = 0, \quad t \in [2, \infty), \quad x_b(t) = 0, t \in [0, \infty) \end{cases}$$

Thus, the solution arrives at the Zeno equilibrium set at the instant $t = 2$. However, this solution is not genuinely Zeno, since $|e_y(t)| = 0 \leq \delta\sqrt{|x_a(t)|}$ holds for any $t \in [0, \infty)$. Therefore, this example shows the necessity of Lemma 1 for Theorem 2.

5. CASE STUDIES

In this section, we provide three case studies to illustrate the feasibility of the proposed results. The first case is about a state and output feedback event-triggered control system with relative triggering conditions. As an application of the sufficient conditions, the second one studies the finite-time event-triggered control, where the state is expected to arrive at the origin in a finite time interval. Our results show that there exists an obstacle to exclude the Zeno phenomenon in this kind of event-triggered control systems. In the third case, using the proposed necessary condition, a triggering condition involving external threshold signals is investigated. Then, the difference of Zeno-freeness and the event-separation property [15], which requires a positive lower bound of inter-event times, is revealed.

5.1. Event-Triggered Control with Relative Triggering Conditions

Consider the following linear time-invariant plant

$$\dot{x} = Ax + Bu + d, \ y = Cx \tag{11}$$

where $A \in \mathbb{R}^{n \times n}$, $B \in \mathbb{R}^{n \times m}$, $C \in \mathbb{R}^{q \times n}$ are constant matrices and (A, B) being controllable. Suppose that A is not Hurwitz, and thus, Assumption 2 holds. If rank$(C) = n$, this is state feedback; otherwise, it denotes the output feedback. The controller is implemented as

$$u(t) = K_y(t_k) = KC_x(t_k), \text{ for } t \in [t_k, t_{k+1}) \tag{12}$$

where $A+BKC$ is supposed to be Hurwitz. Then, one can find matrices $Z > 0$ and $Q > 0$, satisfying

$$(A + BKC)^T Z + Z(A + BKC) = -Q \tag{13}$$

Referring to [15], we design the relative triggering condition as

$$t_{k+1} = \inf\{t \geq t_{kl}\ ||e_y(t)|| > P\,||y(t)||\} \tag{14}$$

where $P > 0$ is to be specified. From [15], if one select $P = \sigma \dfrac{\lambda_{min}(Q)}{2\,||ZBK||\,||C||} > 0$ with $\sigma \in (0, 1)$, then the triggering condition can guarantee Assumption 3 with some $M \geq 0$ that depends on the bound of disturbances. Note that the method in [15] is sufficient and there may exist larger P such that Assumption 3 holds.

Obviously, the Zeno equilibrium set of this event-triggered control system is the kernel space of C (denoted by ker C). Moreover, since the system is linear, the continuity and bounded properties on the functions in (2) and (3) trivially hold. Then, we have the following corollary.

Corollary 3. *Consider the unstable plant in (11) with the controller in (12) and triggering condition in (14). The following statements hold:*

1. *The solution is chattering Zeno if and only if $P < 1$ and there exists a triggering instant $t_{k0} \geq t_0$ such that $x(t_{k0}, k_0) \in \ker C \setminus \Gamma_{d(t)}(t_{k0})$;*
2. *The solution is genuinely Zeno if and only if there exists a triggering instant $t_{k0} > 0$ such that $x(t_{k0}, k_0) \notin \ker C$ and the solution arrives at ker C in a finite time interval.*

Note that the threshold function in (14) is locally compressed if and only if $P < 1$. Thus, Corollary 3 can be directly obtained from Theorems 1 and 2.

Although in [15] it was proved that $P = \sigma \dfrac{\lambda_{min}(Q)}{2\,||ZBK||\,||C||} \in (0,1)$ for any unstable plant in (11), due to the conservatism of their approach, it is not sufficiently excluded that Assumption 3 may hold with $P \geq 1$, which contradicts the local compression of g_0. Therefore, we introduce the following proposition to study when the designers should use a locally compressed threshold function.

Proposition 1. For a given controller gain matrix K and disturbance bound d_{max}, suppose that ker $C \cap \Gamma_x(K, d_{max}) \neq \emptyset$, where $\Gamma_x(K, d_{max})$ is defined in Assumption 2. If the gain matrix K and the triggering condition in (14) is well-designed in the sense of Assumption 3, then necessarily $P < 1$.

Proof. Suppose that the triggering condition is well-designed for $P \geq 1$ to ensure Assumption 3. Since ker $C \cap \Gamma_x(K, d_{max}) \neq \emptyset$, consider an initial states satisfying $x(0, 0) \in$ ker $C \cap \Gamma_x(K, d_{max})$ and $e_y(0, 0) = 0$. Thus, from the fact of $P \geq 1$,

$$\|e_y(t, 0)\| = \|y(t, 0) - y(0, 0)\| = \|y(t, 0)\| \leq P \|y(t_0, 0)\| \qquad (15)$$

hold for all $t \geq 0$. As a result, no events would be triggered any more. Since the triggering condition and the controller is well-designed, the state x is bounded under the constant control signal $u(t) = \kappa(y(0))$ and all disturbance $d(t) \in L^p_\infty$. This conclusion contradicts Assumption 2 since $x(0, 0) \in \Gamma_x(K, d_{max})$. Therefore, the triggering condition in (14) cannot be well-designed if $P \geq 1$ and the proof is completed.

In the case of state feedback, only the origin belongs to the Zeno equilibrium set. If the system is disturbance-free (namely, $d_{max} = 0$), then ker $C = \Gamma_{d(t)}(h) = \{0\}$, $h \in [t_0, \infty)$, and ker $C \cap \Gamma_x(K, 0) = \emptyset$ obviously, which are out of the scope of Proposition 1. Consequently, there may exist $P \geq 1$ such that Assumption 3 holds. On one hand, since ker $C \setminus \Gamma_{d(t)}(h) = \emptyset$ for all $h \in [t_0, \infty)$, there is no chattering Zeno behavior. On the other hand, since the linear controller in (12) can just ensure asymptotic stability, the state x will never arrive at the origin in a finite time interval. Hence, there is no Zeno phenomenon in the considered disturbance-free state feedback event-triggered control systems, which agrees with the results in [3, 15].

In the case of nonzero disturbances or/and output feedback, in general, for unstable A, there always exists an instant $h \geq 0$ such that $\Gamma_{d(t)}(h) \neq$ ker C for some $d(t) \in L^p_\infty$ and $\Gamma_x(K, d_{max}) \cap$ ker $C \neq \emptyset$, which means that only $P < 1$ is admitted from Proposition 1. Thus, the system is Zeno. Compared with [15], a slight contribution of Corollary 3 is to show that it is impossible to exclude genuinely Zeno behavior by just increasing the value of P. Although chattering Zeno behavior may be avoided when $P \geq 1$, it would make the triggering condition in (14) fail to stabilize the unstable plant in most cases of nonzero disturbances or/and output feedbacks.

Then we provide the following simulations to illustrate the feasibility of Proposition 1 and Corollary 3. First we consider the state-feedback systems with/without disturbances.

Example 2. Consider the following plant

$$A = \begin{bmatrix} 0 & 1.8 \\ 0 & -1.8 \end{bmatrix}, B = \begin{bmatrix} 1 \\ 1 \end{bmatrix}, C = \begin{bmatrix} 1 & 0 \\ 0 & 1 \end{bmatrix},$$

and a state-feedback controller gain matrix $K = [-1.8\ 0]$. The eigenvalues of $A+BK$ are $-1.8 + 1.8i$ and $-1.8 - 1.8i$. In this case, we consider the triggering condition in (14) as

$$t_{k+1} = \inf\{t \geq t_k|\ |e_y(t)| > 1.1\ \|y(t)\|\}$$

Note that the parameter $P = 1.1$ is not designed by any existing theoretical methods.

In the absence of disturbance, $P \geq 1$ may be allowable for Assumption 3. Figure 1 shows the state trajectories and the evolution of inter-event times. Obviously, the solution is convergent and Zeno-free. To further illustrate the asymptotic stability of the system, we provide the simulation results under the initial states on the unit circle, i.e., $x(0) = (\sin(\theta_0), \cos(\theta_0))$, $\theta_0 \in [0, 2\pi)$. As shown in Figure 2, all theses solutions are convergent as well. Note that the considered event-triggered control system with (14) is homogeneous (see [33] for more details about the homogeneity of event-triggered control systems). Thus, for two solutions $x_1(t)$ and $x_2(t)$, if $x_1(0) = \alpha x_2(0)$ with some $\alpha \in R$, then $x_1(t) = \alpha x_2(t)$ for all $t \geq 0$. Therefore, the simulations show that the event-triggered control system in this example is asymptotically stable.

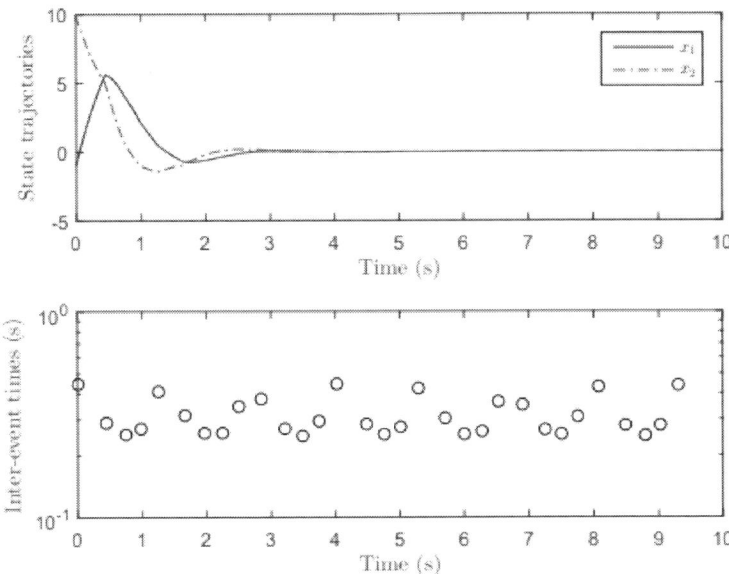

Figure 1. The simulation results of the state feedback event-triggered control system without disturbances.

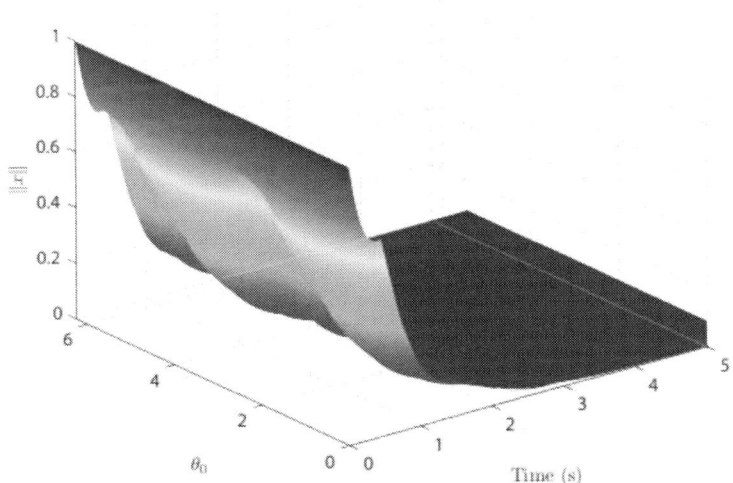

Figure 2. The simulation results of the disturbance-free systems with initial states on unit circle.

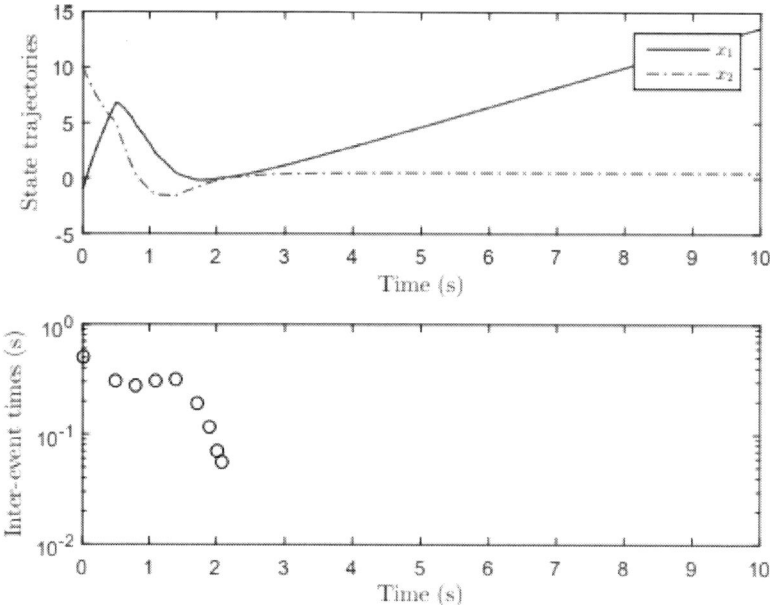

Figure 3. The simulation results of the state feedback event-triggered control system with disturbances.

As a comparison, Figure 3 provides the simulation results for the system under a constant disturbance d = $[1\ 1]^T$, where the state response of the system is divergent. The minimum norm of x(t) is 0.0696 for t ∈ [0, 10s]. Therefore, the state trajectories do not belong to or arrive at the origin in a finite time interval. From the necessity in Corollary 3, the solution is Zeno-free, which agrees with the evolution of inter-event times provided in Figure 3. The simulations show that the linear systems under a relative triggering condition with P ≥ 1 may be asymptotically stable but not input-to-state stable. To the best of our knowledge, this phenomenon is not revealed or discussed in any previous studies. As is well known, for a linear system under a continuous or periodically sampled input, the asymptotic stability in disturbance-free case (namely, internal stability) is always equivalent to its input-to-state stability with respect to additive external disturbances [34].

Thus, this example implies that event-triggered control mechanisms may influence the behavior of a dynamical system in a more essential way than what we used to think. The theoretical analysis on this issue would be left as an open problem, since none of existing theoretical techniques, such as small gain theory [35], can tell their differences.

Zeno behavior in output-feedback control systems is illustrated in the following example.

Example 3. Also consider the plant in Example 2 with the output matrix C = [1 0]. Then, the output-feedback controller gain matrix is K = −1.8. The eigenvalues of A + BKC are −1.8 + 1.8i and −1.8 − 1.8i. An obvious feasible solution of (13) is

$$Z = \begin{bmatrix} 1 & 0 \\ 0 & 1 \end{bmatrix}, Q = \begin{bmatrix} 3.6 & 0 \\ 0 & 3.6 \end{bmatrix},$$

Thus, according to the approach in [15], the relative triggering condition is designed as follows by selecting σ = 0.1414:

$t_{k+1} = \inf\{t \geq t_k | \ |e_y(t)| > 0.1 \ |y(t)|\}.$

Figure 4 provides the state and output trajectories, the evolution of inter-event times, and the number of events when x(0) = [−1 −2]^T and d_{max} = 0. It is shown that the solution of the system is genuinely Zeno. And the Zeno instant is about 1.444s, after which the solution does not exist. The output tends to be zero as the time approaches the Zeno instant. Thus, the simulation results illustrate Corollary 3. Moreover, Figure 4 provides some distinct features of Zeno behavior in simulation, that is, the inter-event times decrease steeply around the Zeno instant and the asymptotic line of the points in the subfigure on Number of events would become a vertical line as *t* approaches the Zeno instant. Thus, by the features, one can recognize Zeno behavior from simulation results.

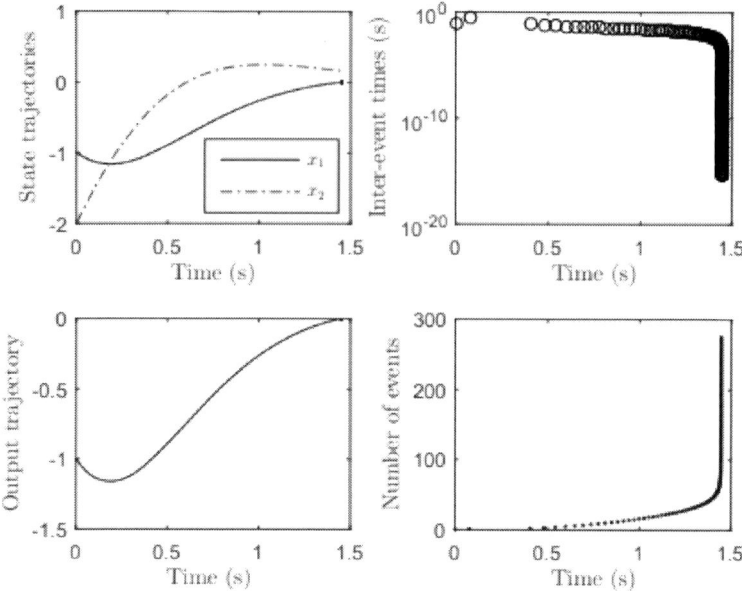

Figure 4. The simulation results of the output feedback event-triggered control system with a relative triggering condition.

5.2. Finite-Time Event-Triggered Control

In this subsection we study the finite-time event-triggered control, where the state is expected to arrive at the origin in a finite time interval. The finite-time stability[1] is defined as follows.

Definition 9. [37] The x-subsystem in (5) is said to be robustly finite-time stable at the origin, if it satisfies the following statements:

1. Robust stability: For any given $\varepsilon > 0$, there exists a constant $\delta(\varepsilon) > 0$ such that if $\max\{\|x(0)\|, d_{\max}\} < \delta(\varepsilon)$, then $\|x(t)\| < \varepsilon$, for $t \in [0, \infty)$.

---

[1] Another interpretation of finite-time stability is to study the control strategy for ensuring some bounded property of states in a finite time interval, see, e.g., [36]. It mainly focuses on the relationship between the upper bound of states within a finite time interval and the bound on initial states. This version of finite-time stability is out of the scope of this subsection.

2. Finite-time convergence: For any initial state $x(0) \in \mathbb{R}^n$, there exists a finite time $T \geq 0$ such that

$$\lim_{t \to T} x(t_k) = 0, and, x(t) = 0, for \quad all \quad t \geq T \qquad (16)$$

The robust stability considered in Definition 9 can be regraded as an improved version of the conventional definition of Lyapunov stability in [34]; and they become the same when $d_0 = 0$.

Obviously, the relative triggering condition introduced in the first subsection cannot guarantee finite-time stability without Zeno behavior since the corresponding state needs to arrive at the origin in a finite time interval, and the origin belongs to the Zeno equilibrium set. In other words, if Zeno behavior is unallowable, there exists no relative event-triggered finite-time control, which is caused by the special forms of threshold functions in relative triggering conditions. Next, we would like to study whether the finite-time event-triggered control can be achieved by other kinds of time-invariant triggering conditions. To this end, we introduce the following definitions and lemmas.

Definition 10. *The threshold function $g_0(g(x))$ in (4) is said to be sufficiently small with respect to the origin if for any $\varepsilon > 0$, there exists a constant $\delta > 0$ such that $g_0(g(x)) < \varepsilon$ for any $x \in \mathbb{R}^n$ satisfying $\|x\| < \delta$.*

By definitions, for example, the relative triggering condition is sufficiently small, and the absolute triggering condition is not sufficiently small.

Lemma 2. *Under Assumption 2, if the plant in (2) with the controller in (3) and the triggering condition in (4) is robustly stable with respect to the origin, then $g_0(g(x))$ is necessarily sufficiently small.*

Proof. We prove this lemma by contradiction. Assume that the closed-loop system is stable in the sense of Definition 9 and the function $g_0(g(x))$ is not sufficiently small. Hence, there exist constants $\eta > 0$ and $\delta_0 > 0$ such that $g_0(g(x)) \geq \eta$ holds for any $x \in \mathbb{R}^n$ satisfying $\|x\| < \delta_0$.

Since the system is robustly stable, for any $L_1 > 0$, there exists $\delta_1 > 0$ such that, for any initial state $x(0)$ and disturbance $d(t)$ with max$\{\|x\|$, $d_{max}\} < \delta_1$, the corresponding solution $x(t)$ satisfies

$$\|x(t)\| \leq \min\{\frac{\eta}{3L_1}, \delta_0\},$$

for all $t \in [0, \infty)$.

Suppose that the Lipschitz constant of $g(x)$ is L_2 in the set $\{x \in \mathbb{R}^n \mid \|x\| \leq \frac{\eta}{3L_1}\}$, and define $L_g = \max\{L_1, L_2\}$. Then similarly, there exists a constant $\delta_g > 0$ such that for any $x(0)$ and $d(t)$ with $\max\{\|x(0)\|, d_{max}\} < \delta_g$, the corresponding solution $x(t)$ satisfies

$$\|x(t)\| \leq \min\{\frac{\eta}{3L_g}, \delta_0\} \leq \min\{\frac{\eta}{3L_1}, \delta_0\},$$

for all $t \in [0, \infty)$. Furthermore, for the above solution $x(t)$, one has that

$$\|e_y(t)\| = \|g(x(t_k)) - g(x(t))\| \leq L_g \|x(t_k) - x(t)\| \leq L_g \frac{2\eta}{3L_g} < \eta,$$

for all $t \in [t_k, t_{k+1})$ and $k \in \mathbb{N}_{\geq 0}$. Thus, $\|e_y(t)\| < \eta \leq g_0(g(x(t)))$ for $t \in [0, \infty)$. This means that the event would not be triggered any more. Define $\delta_2 = \min\{\delta_0, \delta_1, \delta_g\}$. Then, the analysis above implies that the controller $u = \kappa(g(x(0)))$ could lead the state to be bounded for $t \in [0, \infty)$ and $x(0) \in \{x \in \mathbb{R}^n \mid \|x\| < \delta_2\}$, which contradicts Assumption 2. Therefore, the proof is completed.

Lemma 3. *For the time-invariant triggering condition in (4), if $g(g_0(x))$ is sufficiently small, the origin belongs to the corresponding Zeno equilibrium set.*

Proof. This lemma can be easily proved by Definition 10 and the continuity of $g_0(g(x))$.

Based on the lemmas above, we propose the main result in this subsection. It can be regarded as an application of the sufficient condition in Theorem 2 on genuinely Zeno behavior to finite-time event-triggered control systems[2].

Corollary 4. *Under Assumptions 1 and 2 and the triggering condition in (4), no controller can finite-time stabilize the event-triggered control system in (5) without Zeno behavior.*

Proof. We prove this corollary by contradiction. Assume that the closed-loop system is finite-time stable without Zeno behavior under some controller in (3) and time-invariant triggering condition in (4). Then, according to Lemmas 2 and 3, the triggering condition must be sufficiently small with the origin belonging to its Zeno equilibrium set. Moreover, from the converse Lyapunov theorem [34], the robust stability in Definition 9 implies the existence of the function V in Assumption 3.

On one hand, finite-time convergence ensures all the solutions of (5) to arrive at the Zeno equilibrium set (the origin) in a finite time interval. On the other hand, from Assumption 2 and Assumption 3, it follows that the Zeno equilibrium set cannot be equal to the total space, i.e., $\Gamma \neq \mathbf{R}^n$. As a result, there always exists an initial state $x(0) \notin \Gamma$ such that the corresponding solution arrives at the Zeno equilibrium set in a finite time interval. According to Theorem 2, this solution must be genuinely Zeno, and therefore, the proof is completed.

Remark 6. The proof of Corollary 4 depends on only the definition of finite-time stability and the sufficient condition on genuinely Zeno behavior in Theorem 2. Hence, although the proof of Theorem 2 involves the local output equilibrium set $\Gamma_{d(t)}(h)$, it is not necessary to calculate $\Gamma_{d(t)}(h)$. This fact illustrates the analysis in Remark 4 and ensures the results in Corollary 4 to be feasible in a more general sense.

Remark 7. According to the definitions of genuinely Zeno behavior and finite-time stability as well as the proof of Corollary 4, the corresponding solution only exists before the instant T in (16). Hence, strictly speaking, this genuinely Zeno solution cannot really achieve finite-time convergence, since the condition, $x(t) = 0$ for $t \geq T$, cannot be satisfied.

Remark 8. Corollary 4 shows an irreconcilable conflict between finite-time stability and time-invariant event-triggered control. This implies some immense obstacles in the study of finite-time event-triggered control. In fact, none of the current studies on finite-time event-triggered control systems could exclude effectively Zeno behavior if the exact convergence is expected. Hence, to achieve finite-time stability, some more complicated cases might need to be considered, such as, time-varying triggering conditions, time-varying Zeno equilibria, and non-Lipschitz output function g. Note that finite-time convergence is generally achieved by some non-Lipschitz controllers. However, it is still an open problem how to ensure simultaneously a positive lower bound of inter-event times and exact convergence to the equilibrium for systems with non-Lipschitz controllers. Therefore, there remain many challenges to achieve finite-time stability by event-triggered control even if considering these more complicated cases.

Finally, we illustrate the feasibility of Corollary 4 by the following numerical example.

Example 4 Consider the following scalar system:

$$\dot{x}(t) = u(t), u = -\text{sgn}(x)|x(t_k)|^{\frac{1}{2}}$$

which satisfies Assumption 2 obviously. Define the triggering condition as

$$t_{k+1} = \inf\{t \geq t_k | |e(t)| > \frac{1}{2}|x(t)|\},$$

where $e(t) = x(t) - x(t_k)$, $t \in [t_k, t_{k+1})$.

First, we show that the considered closed-loop system satisfies Assumption 3. Define $V = \frac{1}{2}x^2$ then one has

$$\dot{V}(t) = x(t)\dot{x}(t) = -x(t)\sqrt{|x(t_k)|} = -x\sqrt{|x(t) + e(t)|}.$$

Since the triggering condition ensures $|e(t)| \leq \frac{1}{2}|x(t)|$, the signs of $x(t)$ and $x(t) + e(t)$ must be the same for all $t \in [0, \infty)$. Then, for the initial state $x(0) \geq 0$, one has $x(t) \geq 0$, $t \in [0, \infty)$. From the inequalities $\sqrt{|x+e|} \geq \sqrt{|x|} - \sqrt{|e|} > 0$ and $|e| \leq \frac{1}{2}|x|$, it follows that

$$\dot{V}(t) \le -x(t)(\sqrt{|x(t)|} - \sqrt{|e(t)|}) \le \frac{2-\sqrt{2}}{2} V^{\frac{3}{4}}(t) \tag{17}$$

Similarly, one can prove (17) in the case of x(0) < 0. Thus, from Theorem 4.2 in [37], the closed-loop system is finite-time stable (if there is no Zeno behavior) and satisfies Assumption 3 with M = 0.

Next, we calculate the state trajectory and triggering time sequence under the initial state x(0) = 4. Denote by $T_k = t_k - t_{k-1}$, $k \in Z_{\ge 1}$, the inter-event times.

When $t \ge 0$, one has $\dot{x}(t) = -2$, which enforces that x(t) = 4 − 2t and e(t) = 2t before the triggering instant t_1. From the triggering condition in (17), $2T_1 = \frac{1}{2}(4 - 2T_1)$.

As a result, $t_1 = T_1 = \frac{2}{3}$ and $x(t_1) = \frac{8}{3}$.

Then consider $t \ge t_1 = \frac{2}{3}$. We have $\dot{x}(t) = -\sqrt{\frac{8}{3}}$, which yields x(t) = $\frac{8}{3} - \sqrt{\frac{8}{3}}(t-t_1)$ and e(t) = $\sqrt{\frac{8}{3}}(t-t_1)$. Thus, $\sqrt{\frac{8}{3}} T_2 = \frac{1}{2} (\frac{8}{3} - \sqrt{\frac{8}{3}} T_2)$. Consequently, $T_2 = \frac{4}{9}\sqrt{\frac{3}{2}}$, $t_2 = t_1 + T_2 = \frac{2}{3} + \frac{4}{9}\sqrt{\frac{3}{2}}$ and $x(t_2) = \frac{16}{9}$.

Repeating the above processes, one has that

$$T_k = \left(\frac{2}{3}\right)^{\frac{k+1}{2}}, t_k = \sum_{j=1}^{k} T_j = \frac{\frac{2}{3}\left[1-\left(\frac{2}{3}\right)^{\frac{k}{2}}\right]}{1-\sqrt{\frac{2}{3}}} = \frac{2\sqrt{3}\,(1-\sqrt{\frac{2}{3}}^k)}{3\sqrt{3} - 3\sqrt{2}},$$

and $x(t_k) = 4(\frac{2}{3})^k$ for $k \in Z_{\ge 1}$. Obviously, $\lim_{k \to \infty} t_k = t_\infty = \frac{2}{3 - \sqrt{6}} < \infty$, $\lim_{k \to \infty} t_k x(t_k) = 0$ and $T_k > 0$ for all $k \in N_{\ge 1}$. Hence, the corresponding solution is genuinely Zeno, and the origin can be never really arrived at.

5.3. Triggering Conditions with External Threshold Functions

In this subsection, we will study the triggering condition where the triggering function involves an external threshold signal. Specifically, the triggering condition in (4) is replaced by

$$t_{k+1} = \inf\{t \geq t_k | \;\|e_y(t)\| > g_w(y(t), w(t))\} \qquad (18)$$

where $g_w(y, w) := g_0(y) + w$ is the new triggering function and g_0 is the same one as in (4). The auxiliary external signal $w \in \mathbb{R}_{\geq 0}$ is generated by a time-invariant system

$$\dot{w}(t) = -f_1(w(t)), \;and, w(0) \geq 0$$

with some locally Lipschitz continuous function $f_1 : \mathbb{R} \to \mathbb{R}$ satisfying $f_1(0) \geq 0$ and $f_1(s) \leq 0$ for all $s \geq M_w$ and some $M_w \geq 0$. The property of f_1 ensures that $w(t) > 0$ for all $t \in \mathbb{R}_{\geq 0}$ when $w(0) > 0$. Hence, the new Zeno equilibrium set is defined as

$$\Gamma_w := \{[x^T, w]^T \in \mathbb{R}^{n+1} | g_0(g(x)) + w = 0\}.$$

Since the triggering function g_w does not satisfy Assumption 1 and the conditions in Lemma 1 with respect to x, the original sufficient and necessary conditions in Theorem 1 and Theorem 2 do not hold any more. The main reason is that the sufficiency depends on those assumptions. However, by focusing on the necessity of Zeno behavior, one can have the following corollary. The proof is very similar to those in the last section, hence it is omitted.

Corollary 5. *Suppose that Assumption 3 holds for the closed-loop system in (5) with the triggering condition in (18). Then, a Zeno solution χ has to contain an augmented state $[x^T, w]^T$ that either (i) belongs to Γ_w at some triggering instant or (ii) arrives at Γ_w in a finite time interval.*

For further illustration, we consider the following triggering condition proposed in [11]:

$$t_{k+1} = \inf\{t \geq t_k | \;\|e_y(t)\| > c_0 + c_1 e^{-\alpha t}\}, \tag{19}$$

where the constants $c_0 \geq 0$, $c_1 \geq 0$ and $\alpha > 0$ are to be specified. The triggering condition in (19) is equivalent to (18) by selecting $g_0(s) = 0$ and

$$\dot{w}(t) = -\alpha w(t) + \alpha c_0, \text{ and } w(0) = c_0 + c_1.$$

The parameters c_0, $c_1 \geq 0$ are said to be admissible if they satisfy $c_0 + c_1 \neq 0$. Thus, from Corollary 5, one can obtain the following results immediately.

Corollary 6. *Consider the closed-loop system in (5) and the triggering condition in (19). Under Assumption 3, the event-triggered control system is Zeno-free for any admissible parameters c_0 and c_1.*

Proof. Due to $w(t) > 0$ for all $t \geq 0$ under any given admissible parameters, the necessary condition of a Zeno solution in Corollary 5 can never be satisfied, which excludes the possibility of Zeno behavior.

Here, we provide another technical proof to directly show that the accumulation points of triggering time sequences do not exist (i.e., there are only finite transmissions in any finite time interval). The triggering function in (19) implies $\|e_y(t)\| \leq (c_0 + c_1)$ for $t \geq t_0$. According to Assumption 3, the state x is bounded, and subsequently, we have that $\|x(t)\| < x_{max}$ holds for $t \in \mathbb{R}_{\geq 0}$ with some constant $x_{max} > 0$. Next, we consider the derivative of $\|e_y(t)\|$ for $t \in [t_y, t_{k+1})$,

$$\frac{d}{dt}\|e_y(t)\| \leq \|\dot{e}_y(t)\| \leq \left\|\frac{\partial g(x(t))}{\partial x} f(x(t), \kappa(g(x(t)) + e_y(t)), d(t))\right\|.$$

Due to the continuity of f, g, κ and $\frac{\partial g(x(t))}{\partial x}$, the boundedness of $x(t)$, $e_y(t)$ and $d(t)$ implies $\frac{d}{dt}\|e_y(t)\| \leq e_{max}$ with some $e_{max} > 0$. Consequently, by integrating the above differential inequality with $e_y(t_k) = 0$, one has $\|e_y(t)\| \leq e_{max}(t - t_k)$ for all $t \in [t_k, t_{k+1})$.

Consider an arbitrary time interval $[t_a, t_b]$. Without loss of generality, we suppose t_a, $t_b \in \{t_k\}^\infty_{k=0}$. Thus, for this time interval, the minimum inter-event time is lower bounded by $\tau_m(t_b) := \dfrac{c_0 + c_1 e^{-\alpha t_b}}{e_{max}}$. Since c_0 and c_1 are admissible, $\tau_m(t_b) > 0$ for $t_b < \infty$. As a result, the number of transmissions within $[t_a, t_b]$ is upper bounded by $N_m(t_a, t_b) := 1 + \dfrac{(t_b - t_a) e_{max}}{c_0 + c_1 e^{-\alpha t_b}} < \infty$, that is, there are a finite number of transmissions in any finite time interval. Therefore, the proof is completed.

However, as the analysis in [9, 11], to guarantee the event-separation property, some tighter conditions are required on the parameters in (19). Roughly speaking, in the case of $c_0 = 0$ and $c_1 > 0$, the system should be disturbance-free and α needs to be less than the decay rate of the ideal system, where the controller is implemented in a continuous feedback manner, i.e., $u = Ky(t)$ [9]. If we consider a linear plant in (11) and linear controller in (14), from Theorem 3 in [11], α should belong to $(0, |\lambda_{max}(A + BKC)|)$. The above analysis implies that Zeno-free behavior is a broader concept than the event-separation property. Hence, when analyzing the transmission performance, it is not enough to only prove the system to be Zeno-free, see, e.g., [38]. The analysis above is illustrated by the following simulations.

Example 5. *Also consider system in Example 3, where $|\lambda_{max}(A + BKC)| = 1.8$. Then we introduce the following two triggering conditions:*

$$t_{k+1} = \inf\{t \geq t_k | \; |e_y(t)| > e^{-t}\}, \qquad (20)$$

and

$$t_{k+1} = \inf\{t \geq t_k | \; |e_y(t)| > e^{-10t}\}. \qquad (21)$$

Figure 5 shows the simulation results of the system with (20). The minimum inter-event time is 0.0515s. This means that the system has the event-separation property and it is Zeno-free obviously.

Figure 6 provides the simulation results of the system with (21). In the time interval of simulation, the minimum inter-event time is 3.13×10^{-5} s. Although the inter-event times tend to approach zero with t rising up, it is obvious that the slope of the asymptotic line of the points in the 'Number of events' subfigure is not infinitely large. Hence, the system is Zeno-free in simulation. However, the transmission performance of the Zeno-free system in Figure 6 is still dissatisfactory. In other words, the lower bound of inter-event times may be not larger than any positive constant although there always exist finite triggering instants in any finite time interval. Therefore, when designing event-triggered control systems, one also needs to avoid this situation from happening.

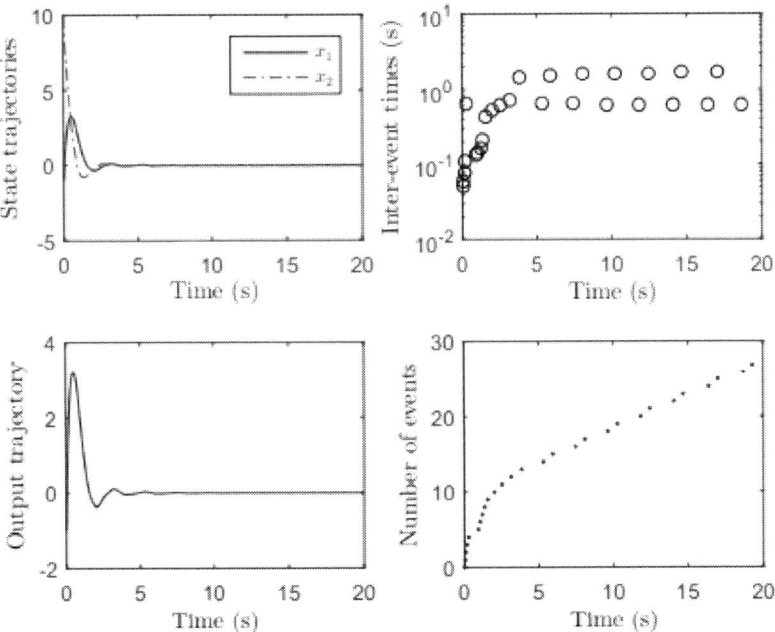

Figure 5. The simulation results of the event-triggered control system with the triggering condition in (20).

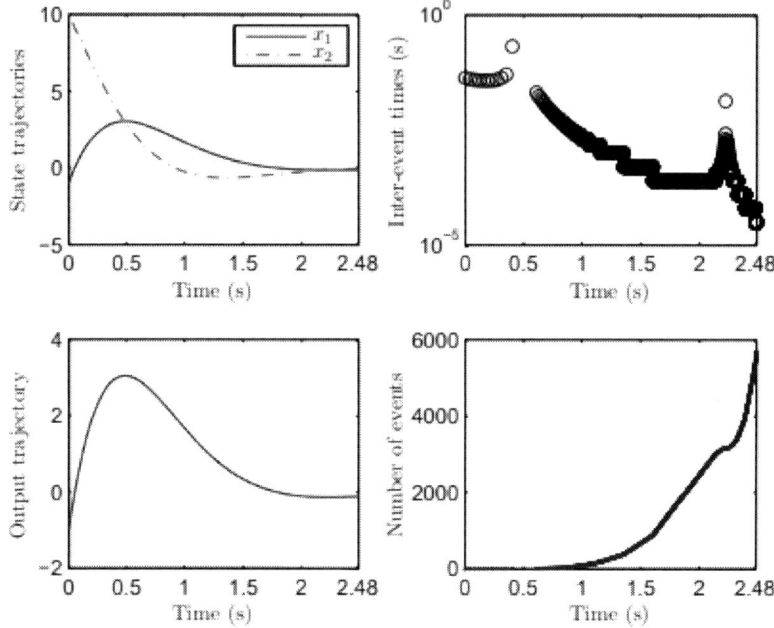

Figure 6. The simulation results of the event-triggered control system with the triggering condition in (21).

CONCLUSION

This chapter has studied the existence of Zeno behavior in event-triggered control systems. Several necessary and sufficient conditions were proposed for the error-based triggering conditions, which were triggered when the difference between the current and the most recently sampled output signals violated a threshold function dependent of output signals. It was shown that Zeno behavior is closely related to finite-time convergence to the Zeno equilibrium set. Then, using the proposed necessary and sufficient conditions, three kinds of event-triggered control systems were investigated in detail. Based on these analyses, it was discovered that an event-triggered control system with a linear plant and controller can be internally stable but not input-to-state stable with respect to external disturbances; some potential conflicts between finite-time stability and time-invariant triggering conditions were revealed; and the difference of Zeno-freeness and the event-separation property was shown. The feasibility of the proposed results was illustrated by numerical examples and simulations. The results in this chapter have some potential extensions. For example, the necessary or the sufficient conditions could be applied to analyze Zeno behavior in the context of distributed event-triggered control [10] or event-triggered consensus control for multi-agent systems [39]. In addition, although our results provided some negative answers to Zeno-free finite-time event-triggered control, this issue may be solved by some more complicated event-triggered control strategies, such as time-varying triggering conditions and non-Lipschitz event output functions, which would be left as future work.

REFERENCES

[1] Åström K. J., Bernhardsson B. M. (1999). Comparison of periodic and event based sampling for first order stochastic systems, *Proceedings of IFAC World Congress*, 11, 301–306.

[2] Årzén K. E., Cervin A., Eker J., Sha L. (2000). An introduction to control and scheduling´ co-design, *Proceedings of IEEE Conference on Decision and Control (CDC)*, 5, 4865–4870.

[3] Tabuada P. (2007). Event-triggered real-time scheduling of stabilizing control tasks, *IEEE Transactions on Automatic Control*, 52(9), 1680–1685.

[4] Wang X., Lemmon M. (2011). On event design in event-triggered feedback systems, *Automatica*, 47(10), 2319–2322.

[5] Garcia E., Antsaklis P. J. (2013). Model-based event-triggered control for systems with quantization and time-varying network delays, *IEEE Transactions on Automatic Control*, 58(2), 422–434.

[6] Sontag E. D. (1989). Smooth stabilization implies coprime factorization, *IEEE transactions on automatic control*, 34(4),435–443.

[7] Donkers M. C. F., Heemels M. (2012). Output-based event-triggered control with guaranteed L_∞-gain and improved and decentralized event-triggering, *IEEE Transactions on Automatic Control*, 57(6), 1362–1376.

[8] Chen X., Hao F. (2013). Observer-based event-triggered control for certain and uncertain linear systems, *IMA Journal of Mathematical Control and Information*, 30(4), 527–542.

[9] Liu T., Jiang Z. (2015). Event-based control of nonlinear systems with partial state and output feedback, *Automatica*, 53, 10–22.

[10] Wang X., Lemmon M. (2011). Event-triggering in distributed networked control systems, *IEEE Transactions on Automatic Control*, 56(3), 586–601.

[11] Guinaldo M., Dimarogonas D. V., Johansson K. H., Sanchez J.., Dormido S. (2013). Distributed event-based control strategies for interconnected linear systems, *IET Control Theory & Applications*, 7(6), 877–886.

[12] Eqtami A., Dimarogonas D. V., Kyriakopoulos K. J. (2010). Event-triggered control for discrete-time systems, *Proceedings of American Control Conference (ACC)*, 4719–4724.

[13] Heemels M., Donkers M. C. F., Teel A. R. (2013). Periodic event-triggered control for linear systems, *IEEE Transactions on Automatic Control*, 58(4), 847–861.

[14] Lunze J., Lehmann D. (2010). A state-feedback approach to event-based control, *Automatica*, 46(1), 211–215.

[15] Borgers D. P., Heemels M. (2014). Event-separation properties of event-triggered control systems, *IEEE Transactions on Automatic Control*, 59(10), 2644–2656.

[16] Goebel R., Sanfelice S., Teel A. R. (2009). Hybrid dynamical systems, *IEEE Control Systems Magazine*, 29(2), 28–93.

[17] Liberzon D., NeŠiĆ D., Teel A. R. (2014). Lyapunov-based small-gain theorems for ´ hybrid systems, *IEEE Transactions on Automatic Control*, 59(6), 1395– 1410.

[18] Abdelrahim M., Postoyan R., Daafouz J., NeŠiĆ D. (2017). Robust event-triggered out- put feedback controllers for nonlinear systems, *Automatica*, 45, 96–108.

[19] Dolk V., Heemels M. (2017). Event-triggered control systems under packet losses, *Automatica*, 80, 143–155.

[20] Postoyan R., Tabuada P., Nesič D., Anta A. (2015). A framework for the event-triggered stabilization of nonlinear systems, *IEEE Transactions on Automatic Control*, 60(4), 982–996.

[21] Yu H., Hao F., Chen T. (2019). A uniform analysis on input-to-state stability of decentralized event-triggered control systems, *IEEE Transactions on Automatic Control*, 64(8), 3423–3430.

[22] Heymann M., Feng L., Meyer G. (1998). Synthesis and viability of minimally interventive legal controllers for hybrid systems, *Discrete Event Dynamic Systems*, 8(2), 105–135.

[23] Ames A. D., Abate A., Sastry S. (2005). Sufficient conditions for the existence of Zeno behavior, *Proceedings of IEEE Conference on Decision and Control and European Control Conference (CDC-ECC)*, 696–701.

[24] Lamperski A., Ames A. D. (2007). Lyapunov-like conditions for the existence of Zeno behavior in hybrid and Lagrangian hybrid systems, *Proceedings of IEEE Conference on Decision and Control (CDC)*, 115–120.

[25] Goebel R., Teel A. R. (2008). Lyapunov characterization of Zeno behavior in hybrid systems, *Proceedings of IEEE Conference on Decision and Control (CDC)*, 2752–2757.

[26] Or Y., Ames A. D. (2011). Stability and completion of Zeno equilibria in Lagrangian hybrid systems, *IEEE Transactions on Automatic Control*, 56(6), 1322–1336.

[27] Lamperski A., Ames A. D. (2013). Lyapunov theory for Zeno stability, *IEEE Transctions on Automatic Control*, 58(1), 100–112.

[28] Hong Y., Xu Y., Huang J. (2002). Finite-time control for robot manipulators, *System & Control Letters*, 46(4), 243–253.

[29] Yu S., Wang Y., Jin L. (2017). Comments on "Finite-time consensus of multi-agent system via non-linear event-triggered control strategy," *IET Control Theory & Applications*, 11(10), 1658–1661.

[30] Zhu Y., Guan X., Luo X., Li S. (2015). Finite-time consensus of multi-agent system via nonlinear event-triggered control strategy, *IET Control Theory & Applications*, 9(17), 2548–2552.

[31] Zhang J., Johansson K. H., Lygeros J., Sastry S. (2001). Zeno hybrid systems, *International Journal of Robust and Nonlinear Control*, 11(5), 435–451.

[32] Yu H., Hao F. (2020). The existence of Zeno behavior and its application to finitetime event-triggered control, *Science China Information Sciences*, 63(1), 139201.

[33] Anta A., Tabuada P. (2010). To sample or not to sample: Self-triggered control for nonlinear systems, *IEEE Transactions on Automatic Control*, 55(9), 2030–2042.

[34] Khalil H. (2002). *Nonlinear Systems Upper Saddle River*, NJ: Prentice Hall.

[35] Liu T., Jiang Z. P. (2015). A small-gain approach to robust event-triggered control of nonlinear systems, *IEEE Transactions on Automatic Control*, 60(8), 2072–2085.

[36] Amato F., Ariola M., Dorato P. (2001). Finite-time control of linear systems subject to parametric uncertainties and disturbances, *Automatica*, 37(9). 1459–1463.

[37] Bhat S. P., Bernstein D. S. (2000). Finite-time stability of continuous autonomous systems, *Siam Journal on Control & Optimization,* 38(3), 751–766.

[38] Li H., Liao X., Huang T., Zhu W. (2015). Event-triggering sampling based leaderfollowing consensus in second-order multi-agent systems, *IEEE Transactions on Automatic Control*, 60(7), 1998–2003.

[39] Zhang Z., Zhang L., Hao F., Wang L. (2017). Leader-following consensus for linear and Lipschitz nonlinear multiagent systems with quantized communication, *IEEE Transactions on Cybernetics*, 47(8), 1970–1982.

In: Networked Control Systems
Editors: S. Tong and D. Qian
ISBN: 978-1-53619-892-8
© 2021 Nova Science Publishers, Inc.

Chapter 6

ROBUST GUARANTEED PERFORMANCE CONSENSUS FOR MAS WITH TIME-DELAYS AND UNCERTAINTY

Yaxiao Zhang[1,*] *and Shiwen Tong*[1,2]

[1]College of Robotics, Beijing Union University, Beijing, China
[2]State Key Laboratory for Management and Control of Complex Systems, Institute of Automation, Chinese Academy of Sciences, Beijing, China

ABSTRACT

In this chapter a robust guaranteed performance consensus problem for continuous-time linear high-order multi-agent systems with uncertainties and time-varying delays is studied. Firstly, the robust guaranteed performance consensus problem is transformed into a robust guaranteed performance control problem of an auxiliary uncertain system with time-varying delays by a linear transformation. Secondly, a sufficient condition for the robust guaranteed performance consensus problem is presented in terms of linear matrix inequality techniques by robust guaranteed performance control theory, and an upper bound of the guaranteed performance function is given. Finally, a numerical example is shown to demonstrate the above theoretical results.

---

* Corresponding Author's E-mail: zdhtyaxiao@buu.edu.cn.

Keywords: multi-agent systems; uncertain systems; robust guaranteed performance; consensus; time-varying delays

1. INTRODUCTION

In recent years, the problem of distributed coordinated control of multi-agent system has been paid much attention by researchers, and has been widely used in formation, synchronization, capture and other practical systems. Consensus problem has become one of the basic problems in cooperative control of multi-agent systems, that is to design appropriate cooperative control protocol so that a certain state of all agents can reach a common value or the same dynamic process.

At present, abundant achievements have been made in the research of multi-agent system consensus. In literature [1], the consensus problem of multi-agent systems with a directed fixed communication topology was studied. However, in practical applications, information transmission between agents is carried out through communication devices. Due to the limited transmission speed, channel congestion and packet loss in the communication process, communication delay will be caused, which will have a certain impact on the whole system. In literature [2], the consensus problem of high order continuous-time linear multi-agent system with time-varying communication delay protocol was studied. The consensus problem was equivalent to the asymptotic stability problem of the corresponding time-delay system by using linear transformation and state space decomposition. In Literature [3], the robust consensus problem was transformed into the robust stability problem of the corresponding system by using linear transformation, and the sufficient criterion of robust consensus in the form of LMIS was obtained.

The traditional research just focuses on the stability of the system, does not consider the performance index of the system, such as the energy consumption of the system. However, in practical applications, each agent may have a limited energy supply to perform certain tasks.

At the same time, in order to meet the requirements of some performance indicators, consensus adjustment is needed. How to achieve the trade-off design between consensus regulation performance and energy consumption is a key challenge, which has a very important research value. Literature [4] first proposed the guaranteed performance consensus problem, and literature [5, 6] studied the first-order guaranteed performance consensus problem of continuous time multi-agent system. Literature [7] studied the second-order guaranteed performance consensus problem of continuous time multi-agent systems. In literature [8, 9], the problem of robust guaranteed performance consensus for high-order discrete-time uncertain multi-agent system was considered. However, the problem of robust guaranteed performance consensus with communication time delay and topology uncertainty was rarely considered. Inspired by the literature mentioned above, this chapter studies the robust guaranteed performance consensus problem of high-order linear continuous time multi-agent systems with communication delays and uncertain topologies.

The rest of the chapter is organized as follows: Section 2 shows the problem description based on graph theory. Section 3 presents a linear transformation approach to transform the robust guaranteed cost consensus problem into a robust stability problem of a corresponding auxiliary system and derives a sufficient condition for robust guaranteed-cost consensus, and the upper bound of guaranteed cost are presented. Numerical results are presented in Section 4, and then concluding remarks are given.

Notations: R^n and $R^{n \times m}$ are the n-dimension real column vector and the set of $n \times m$ dimensional real matrices, respectively. Let 0 be zero number, zero vectors, or zero matrices in appropriate dimension, respectively. Let 1_N denote an N-dimensional column vector with 1. Let P^T and P^{-1} denote the transpose and the inverse matrix of P, respectively. $P^T = P > 0$ stands for matrix P is symmetric and positive definite. The notation * denotes the symmetric terms of a symmetric matrix. $\otimes$ is the Kronecker product of matrices.

2. PROBLEM STATEMENT

Consider a Linear multi-agent system consisting of N agents, where each agent takes the following dynamics:

$$\dot{x}_i(t) = Ax_i(t) + Bu_i(t), \ x_i(0) = x_{i0}, \ i \in I \tag{1}$$

where $x_i \in R^n$ is the state of agent, $A \in R^{n \times n}, B \in R^{n \times m}$ are system matrix and input matrix (A, B) are stabilizable, $u_i \in R^m$ is consensus protocol, which depends on the relative state $(x_j - x_i)$, agent j is called a neighbor of agent i if there exists a communication channel from j to i finite set $I = \{1, \cdots, N\}$. Denote $N_i(t)$ be the set of the neighbors of the agent i at time t and $\mathcal{N}(t) =: \{N_i(t), i \in \mathcal{I}\}$ be a communication configuration of the system (1) at time. $\mathcal{N}(t)$ can be expressed by a digraph $G = \{V, E(t), W(t)\}$. Vertex set $V = \{1, \cdots, N\}$ represents the group of agents, time-varying edge set $E(t) \subseteq V \times V$ denotes the communication topology $\mathcal{N}(t)$, i.e., $(j, i) \in E(t) \Leftrightarrow j \in N_i(t)$, and $W(t) = [\omega_{ij}] \in R^{N \times N}$ is a weighted adjacency matrix, which represents communication connection structure. Consensus protocol with time-varying communication delay and uncertain constraints is considered as follows:

$$u_i(t) = K \sum_{j \in N_i(t)} (\omega_{ij} - \Delta\omega_{ij}(t))[x_j(t - d(t)) - x_i(t - d(t))] \tag{2}$$

where K is the gain matrix, $\Delta\omega_{ij}(t)$ represents the uncertainty of communication topology, $\Delta\omega_{ii}(t) = 0$, $d(t)$ is the time-varying communication delay, satisfies the following conditions:

Assumption 1 $d(t)$ is a differentiable function which satisfies $0 \leq d(t) \leq \tau$ and $|\dot{d}(t)| \leq \delta < 1$, where τ and δ are known constant.

By substituting the protocol (2) into the multi-agent system (1), we can obtain:

$$\begin{cases} \dot{x}(t) = (I_N \otimes A)x(t) - [(L - \Delta L) \otimes BK]x(t - d(t)) \\ x(\theta) = \varphi(\theta), \theta \in [-\tau, 0] \end{cases} \quad (3)$$

where $x = [x_1^T \cdots x_N^T]^T \in R^{Nn}$, $\varphi(\theta)$ is a continuously differentiable initial condition function. $L = [l_{ij}]$ is a weighting Laplacian matrix that derived from communication topology N:

$$l_{ij} = \begin{cases} \sum_{k \in N_i} \omega_{ik}(t), j = i \\ -\omega_{ij}(t), j \neq i, j \in N_i(t). \\ 0, j \notin N_i(t) \end{cases}$$

Similarly, define uncertainty Laplacian matrix $\Delta L(t) = [\Delta l_{ij}(t)]$. There are assumptions for communication topology N and uncertainty Laplacian matrix $\Delta L(t)$ as follows:

Assumption 2 There exist a spanning tree for communication topology N.

Assumption 3 It can be expressed in the following form for the uncertainty Laplacian matrix $\Delta L(t)$:

$$\Delta L(t) = L^1 \mathrm{L}(t) L^2, \quad (4)$$

where L^1 and L^2 are constant matrices of proper dimensions, $\mathrm{L}(t)$ is an unknown time varying matrix which satisfies

$$\mathrm{L}^T(t)\mathrm{L}(t) \leq I. \quad (5)$$

Consider the following linear quadratic cost function:

$$J_C = \int_0^\infty \sum_{i=1}^N u_i^T(t) Z_u u_i(t) dt + \int_0^\infty \sum_{i=1}^N \sum_{j=1}^N \omega_{ij} [(x_j(t) - x_i(t))^T Z_x (x_j(t) - x_i(t))] dt, \quad (6)$$

where $Z_u \in R^{m \times m}$, $Z_x \in R^{n \times n}$ are given symmetric positive definite matrices.

Definition 1 Linear multi-agent system(LMAS) (1) is said to achieve consensus robustly via the consensus protocol (2)with time-varying communication delay and uncertainty, if for any initial state variable $\varphi(\theta)$ there exist a sequence $\xi(t)$ and a positive number J_C^*, such that there are $\lim_{t \to \infty} \|x_i(t) - \xi(t)\| = 0$ and $J_C \leq J_C^*$, and $\xi(t)$ is called the state consensus function, and J_C^* is called a guaranteed cost.

Next, we equivalently transform the robust consensus problem of a multi-agent system with time-varying communication delay and uncertain communication topologies into an robust guaranteed cost control problem of a corresponding auxiliary uncertain systems by a linear transformation, and analyze the sufficient condition that system (1). The robust guaranteed cost consensus criterion, state consensus function and the upper bound of guaranteed cost function for a high order linear multi-agent system (1) with time-varying communication delay and uncertainty protocol (2) are obtained.

3. ROBUST GUARANTEED COST CONSENSUS ANALYSIS

3.1. Problem Transformation

Transform system (3) by the following linear transformation:

$$\bar{x}(t) = Sx, \tag{7}$$

where

$$S = \begin{bmatrix} \tilde{S}_0 \\ 1_N^T \end{bmatrix} \otimes I_n, \ \tilde{S}_0 = \begin{bmatrix} 1 & -1 & 0 & \cdots & 0 \\ 0 & 1 & -1 & \cdots & 0 \\ \vdots & \ddots & \ddots & \ddots & \vdots \\ 0 & \cdots & 0 & 1 & -1 \end{bmatrix},$$

The inverse matrix of S^{-1} can be worked out as follows

$$\tilde{S}_0 = \frac{1}{N}\begin{bmatrix} N-1 & N-2 & \cdots & 1 & 1 \\ -1 & N-2 & \cdots & 1 & 1 \\ \vdots & \vdots & \ddots & \vdots & \vdots \\ -1 & -2 & \cdots & 1 & 1 \\ -1 & -2 & \cdots & -(N-1) & 1 \end{bmatrix} \otimes I_n := \begin{bmatrix} \hat{S}_0 & N^{-1}1_N \end{bmatrix} \otimes I_n.$$

By the linear transformation (7), system (3) is transformed into the following system:

$$\begin{cases} \dot{\bar{x}}(t) = S(I_N \otimes A)S^{-1}\bar{x}(t) - S[(L - \Delta L(t)) \otimes BK]S^{-1}\bar{x}(t-d(t)), \\ \bar{x}(\theta) = \bar{\varphi}(\theta), \theta \in [-\tau, 0], \end{cases} \quad (8)$$

where $\bar{\varphi}(\theta) = S\varphi(\theta)$.

Let $\bar{x} = [y^T \ z^T]^T$, where $y = [\bar{x}_1^T \cdots \bar{x}_{N-1}^T]^T$, $z = \bar{x}_N$. Introduce the definition of the robust stability of the equilibrium point $\bar{x} = 0$ for system (8) with respect to partial variables y.

Definition 2. The equilibrium point $\bar{x} = 0$ of the system (8) is said to be asymptotic stable with respect to the partial variables y (or briefly robust y-stable) if for any $\varepsilon > 0$, there exist a $\delta(\varepsilon) > 0$, such that state trajectory $\bar{x}(t) = [y^T(t) \ z^T(t)]$ satisfies $\|\bar{y}(\theta)\| \leq \delta(\varepsilon)$ and $z(\theta) \in R^n$, there is $\sup_{t \geq 0}\|y(\theta)\| < \varepsilon$, $\lim_{t \to \infty}\|y(t)\| = 0$ for any initial state function $\bar{x}(\theta), \theta \in [-\tau, 0]$.

Thus, the following necessary and sufficient conditions can be obtained:

Theorem 1. LMAS (1) is said to achieve consensus robustly under the protocol (2), if for any initial condition sequence $x(0)$, if and only if the following systems with structural uncertainty and time delay constraints

$$\dot{y}(t) = \bar{A}y(t) + (\bar{B}\bar{K} + H\bar{L}(t)E\bar{K})y(t-d(t)) \quad (9)$$

is robust stable.

Where $\bar{A} = I_{N-1} \otimes A$, $\bar{B} = -(\tilde{S}_0 L \hat{S}_0) \otimes B$, $\Delta \bar{B} = (\tilde{S}_0 \Delta L(t) \hat{S}_0) \otimes B = H \bar{L}(t) E$,
$\bar{L}^T(t)\bar{L}(t) \leq I$. $\bar{K} = I_{N-1} \otimes K$, $\bar{C} = -[1_N^T(L - \Delta L(t))\hat{S}_0] \otimes BK$.

Proof: Firstly, through the protocol (2), the multi-agent system (1) can be expressed as system (3). By linear transformation (7), system (3) is equivalent to system (8), it knows $\lim_{t \to \infty} \|x_i(t) - x_j(t)\| = 0$, $i = 1, \cdots, N$ if and only if $\lim_{t \to \infty} \|y_i(t)\| = 0$, $i = 1, \cdots, N-1$, through the protocol (2), LMAS (1) can achieve asymptotically robust if and only if system (8) is robust y-stable.

Now expand system (8), and there is:

$$\dot{x}(t) = \begin{bmatrix} \dot{y}(t) \\ \dot{z}(t) \end{bmatrix} = (\begin{bmatrix} \tilde{S}_0 \\ 1_N^T \end{bmatrix} \otimes I_n)(I_N \otimes A)(\begin{bmatrix} \hat{S}_0 & N^{-1}1_N \end{bmatrix} \otimes I_n) \begin{bmatrix} y(t) \\ z(t) \end{bmatrix}$$
$$-(\begin{bmatrix} \tilde{S}_0 \\ 1_N^T \end{bmatrix} \otimes I_n)[(L - \Delta L(t)) \otimes BK](\begin{bmatrix} \hat{S}_0 & N^{-1}1_N \end{bmatrix} \otimes I_n) \begin{bmatrix} y(t-d(t)) \\ z(t-d(t)) \end{bmatrix}$$
$$= \begin{bmatrix} I_{N-1} \otimes A & 0 \\ 0 & A \end{bmatrix} \begin{bmatrix} y(t) \\ z(t) \end{bmatrix} - \begin{bmatrix} [\tilde{S}_0(L - \Delta L(t))\hat{S}_0] \otimes BK & 0 \\ [1_N^T(L - \Delta L(t))\hat{S}_0] \otimes BK & 0 \end{bmatrix} \begin{bmatrix} y(t-d(t)) \\ z(t-d(t)) \end{bmatrix}.$$
$$= \begin{bmatrix} \bar{A} & 0 \\ 0 & A \end{bmatrix} \begin{bmatrix} y(t) \\ z(t) \end{bmatrix} + \begin{bmatrix} (\bar{B} + \Delta \bar{B})\bar{K} & 0 \\ \bar{C} & 0 \end{bmatrix} \begin{bmatrix} y(t-d(t)) \\ z(t-d(t)) \end{bmatrix}.$$

Therefore, system (8) is equivalent to

$$\begin{cases} \dot{y}(t) = \bar{A}y(t) + (\bar{B} + \Delta\bar{B})\bar{K}y(t-d(t)) \\ \dot{z}(t) = Az(t) + \bar{C}z(t-d(t)) \end{cases} \quad (10)$$

As $\dot{y}(t)$ in the first equation of system (10) (e.g., system (9)) is independent to $z(t)$ and $z(t-d(t))$, Therefore, the robust stability of system (8) is equivalent to the robust stability of system (9).

Therefore, by protocol (2), the multi-agent system (1) can achieve asymptotic robust consensus if and only if the system (9) is robust stable.

Guaranteed function (6) can be written as

$$J_C = \int_0^\infty y^T(t)\{2(L-\Delta L) \otimes Z_x + [(L-\Delta L)^T(L-\Delta L)] \otimes K^T Z_u K\} y(t) dt \quad (11)$$

According to Theorem 1 and Definition 1, we can get:

Theorem 2. LMAS (1) is said to achieve guaranteed cost consensus robustly under the protocol (2), if and only if system (9) is robust stable. and there exists a $J_C^* > 0$, such that $J_C \leq J_C^*$.

3.2. Consensus Criterion

Lemma 1 Schur Complement [10]: For given symmetric matrix

$$P = P^T = \begin{bmatrix} P_{11} & P_{12} \\ * & P_{22} \end{bmatrix}$$

where $P_{11} \in R^{m \times m}, P_{22} \in R^{n \times n}$, Then the following three conditions are equivalent:

$P < 0$

$P_{11} < 0, P_{22} - P_{12}^T P_{11}^{-1} P_{12} < 0;$

$P_{22} < 0, P_{11} - P_{12}^T P_{22}^{-1} P_{12} < 0.$

Lemma 2. [11] For given matrices $Q = Q^T, H, E$, there is

$$Q + HL(t)E + E^T L^T(t) H^T < 0,$$

for any $L(t)$ that satisfies (5) if and only if there exist a $\lambda > 0$ such that

$$Q + \lambda^{-1} HH^T + \lambda E^T E < 0.$$

Theorem 3 Suppose Assumption 1 hold, LMAS (1). LMAS (1) is said to achieve guaranteed cost consensus robustly under the protocol (2), for given $\tau > 0$ and δ, f there exist n-dimension matrices

W, $P = P^T > 0$, $n(N-1)$-dimension matrices $Q = Q^T \geq 0$, $R = R^T \geq 0$, any matrices of appropriate dimensions $X = \begin{bmatrix} X_{11} & X_{12} \\ * & X_{22} \end{bmatrix} \geq 0$, Y_1 and Y_2 and a scalar $\lambda > 0$, such that, for system (9), the following LMI is valid:

$$\Phi = \begin{bmatrix} \phi_{11} & \phi_{12} & \tau \bar{A} R & PH \\ * & \phi_{22} + \lambda E^T E & \tau \bar{B} \bar{K} R & 0 \\ * & * & -\tau R & \tau RH \\ * & * & * & -\lambda I \end{bmatrix} < 0, \quad (12)$$

$$\Psi = \begin{bmatrix} X_{11} & X_{12} & Y_1 \\ * & X_{22} & Y_2 \\ * & * & R \end{bmatrix} \geq 0, \quad (13)$$

where

$$\begin{cases} \phi_{11} = \mathbf{Y} + Y_1 + Y_1^T + \tau X_{11}, \\ \phi_{12} = (I_{N-1} \otimes P)\bar{B}(I_{N-1} \otimes W) + [\bar{B}(I_{N-1} \otimes W)]^T (I_{N-1} \otimes P) - Y_1 + Y_2^T + \tau X_{12}, \\ \phi_{22} = -Y_2 - Y_2^T - (1-\delta)Q + \tau X_{22}, \\ \mathbf{Y} = 2L \otimes Z_x + (L^T L) \otimes (K^T Z_u K) + (I_{N-1} \otimes P)\bar{A} + \bar{A}^T (I_{N-1} \otimes P) + Q. \end{cases} \quad (14)$$

The control gain matrix $K = WP$ in protocol (2), and the upper bound of corresponding system performance is

$$J_C^* = y_0^T (I_{N-1} \otimes P) y_0 + \tau y_0^T (Q + R) y_0.$$

Proof: According to Theorem 2, it is only necessary to prove that system (9) is robust stable.

The following Lyapunov-Krasovskii functional is selected for system (9):

$$V(t,y(t)) = y^T(t)(I_{N-1} \otimes P)y(t) + \int_{t-d(t)}^{t} y^T(s)Qy(s)ds + \int_{-\tau}^{0}\int_{t+\theta}^{t} \dot{y}^T(s)R\dot{y}(s)dsd\theta.$$

where, $P = P^T > 0$, $Q = Q^T \geq 0$, $R = R^T \geq 0$ are undetermined matrices. let $\eta_1 = [y^T(t) \ y^T(t-d(t))]^T$, for any positive definite matrices in appropriate dimension $X = \begin{bmatrix} X_{11} & X_{12} \\ * & X_{22} \end{bmatrix}$ there is

$$\tau \eta_1^T(t) X \eta_1(t) - \int_{t-d(t)}^{t} \eta_1^T(t) X \eta_1(t)ds \geq 0. \tag{15}$$

And according to Neoton-Leibneiz formula, there is

$$y(t-d(t)) = y(t) - \int_{t-d(t)}^{t} \dot{y}(s)ds$$

Hence, for any free weight matrix of appropriate dimensions Y_1 and Y_2, there is

$$2[y^T(t)Y_1 + y^T(t-d(t))Y_2][y(t) - \int_{t-d(t)}^{t} \dot{y}(s)ds - y(t-d(t))] = 0 \tag{16}$$

Take the derivative of the functional $V(t,y(t))$ along system (9), let $\overline{P} = I_{N-1} \otimes P$, it draws

$$\dot{V}(t,y(t)) = y^T(t)[\overline{P}\overline{A} + \overline{A}^T \overline{P} + Q]y(t) + 2y^T(t)\overline{P}(\overline{BK} + H\overline{L}(t)E\overline{K})^T y(t-d(t))$$
$$-(1-\dot{d}(t))y^T(t-d(t))Q y(t-d(t)) + \tau \dot{y}^T(t)R\dot{y}(t) - \int_{t-\tau}^{t} \dot{y}^T(s)R\dot{y}(s)ds.$$

Defining Functions

$$J(t) = \dot{V}(t,y(t)) + \overline{J}_C, \tag{17}$$

where $\overline{J}_C \geq 0$

and

$$\bar{J}_C = y^T(t)\{2(L-\Delta L)\otimes Z_x + [(L-\Delta L)^T(L-\Delta L)]\otimes K^T Z_u K\}y(t).$$

Substitute (15) and the left expressions of (16), therefore, there is

$$\dot{J}(t) \leq y^T(t)[\bar{P}\bar{A}+\bar{A}^T\bar{P}+Q+2(L-\Delta L)\otimes Z_x + [(L-\Delta L)^T(L-\Delta L)]\otimes K^T Z_u K]y(t)$$
$$+2y^T(t)\bar{P}(\bar{B}\bar{K}+H\bar{L}(t)E\bar{K})^T y(t-d(t)) - (1-\dot{d}(t))y^T(t-d(t))Q y(t-d(t))$$
$$+\tau \dot{y}^T(t)R\dot{y}(t) - \int_{t-\tau}^{t}\dot{y}^T(s)R\dot{y}(s)ds + 2[y^T(t)Y_1 + y^T(t-d(t))Y_2][y(t) - \int_{t-d(t)}^{t}\dot{y}(s)ds - y(t-d(t))]$$
$$+\tau \eta_1^T(t)X\eta_1(t) - \int_{t-d(t)}^{t}\eta_1^T(t)X\eta_1(t)ds$$

Let

$$Y = 2L\otimes Z_x + (L^T L)\otimes(K^T Z_u K) + (I_{N-1}\otimes P)\bar{A} + \bar{A}^T(I_{N-1}\otimes P) + Q,$$

There is

$$\dot{J}(t) \leq y^T(t)Y y(t) + 2y^T(t)\bar{P}(\bar{B}\bar{K}+H\bar{L}(t)E\bar{K})^T y(t-d(t))$$
$$-(1-\dot{d}(t))y^T(t-d(t))Q y(t-d(t)) + \tau \dot{y}^T(t)R\dot{y}(t) - \int_{t-\tau}^{t}\dot{y}^T(s)R\dot{y}(s)ds$$
$$+2[y^T(t)Y_1 + y^T(t-d(t))Y_2][y(t) - \int_{t-d(t)}^{t}\dot{y}(s)ds - y(t-d(t))] + \tau \eta_1^T(t)X\eta_1(t)$$
$$-\int_{t-d(t)}^{t}\eta_1^T(t)X\eta_1(t)ds - y^T(t)[2\Delta L\otimes Z_x + (L^T\Delta L + \Delta L^T L - \Delta L^T\Delta L)\otimes K^T Z_u K]y(t)$$
$$< y^T(t)Y y(t) + 2y^T(t)\bar{P}(\bar{B}\bar{K}+H\bar{L}(t)E\bar{K})^T y(t-d(t))$$
$$-(1-\dot{d}(t))y^T(t-d(t))Q y(t-d(t)) + \tau \dot{y}^T(t)R\dot{y}(t) - \int_{t-\tau}^{t}\dot{y}^T(s)R\dot{y}(s)ds$$
$$+2[y^T(t)Y_1 + y^T(t-d(t))Y_2][y(t) - \int_{t-d(t)}^{t}\dot{y}(s)ds - y(t-d(t))]$$
$$+\tau \eta_1^T(t)X\eta_1(t) - \int_{t-d(t)}^{t}\eta_1^T(t)X\eta_1(t)ds$$
$$=\eta_1^T(t)\Gamma\eta_1(t) - \int_{t-d(t)}^{t}\eta_2^T(t,s)\Psi\eta_2(t,s)ds.$$

where

$$\eta_2(t,s) = [y^T(t) \ y^T(t-d(t)) \ \dot{y}^T(s)]^T,$$

$$\Gamma = \begin{bmatrix} \phi_{11} + \tau \bar{A}^T R \bar{A} & \gamma_{12} \\ * & \gamma_{22} \end{bmatrix},$$

$$\gamma_{12} = \phi_{12} + \tau \bar{A}^T R (\bar{B}\bar{K} + H\bar{L}(t)E\bar{K}),$$

$$\gamma_{22} = \phi_{22} + \tau (\bar{B}\bar{K} + H\bar{L}(t)E\bar{K})^T R (\bar{B}\bar{K} + H\bar{L}(t)E\bar{K}).$$

And Ψ is defined in Equation (13), ϕ_{11}, ϕ_{12} and ϕ_{22} are defined in Equation (14). If $\Gamma < 0$ and $\Psi \geq 0$, thus, for sufficiently small ε, there is $\dot{V}(t, y(t)) < -\varepsilon \|y(t)\|^2$, in other words, the robust stability of the system (9) under the constraints of time-varying communication delays and uncertainties can be guaranteed. Apply Lemma 1, $\Gamma < 0$ is equivalent to

$$\begin{bmatrix} \phi_{11} & \phi_{12} & \tau \bar{A}^T R \\ * & X_{22} & \tau(\bar{B}\bar{K})^T R \\ * & * & -\tau R \end{bmatrix} + \begin{bmatrix} \bar{P}H \\ 0 \\ \tau RH \end{bmatrix} \bar{L}(t) \begin{bmatrix} 0 & E & 0 \end{bmatrix} + \begin{bmatrix} 0 \\ E^T \\ 0 \end{bmatrix} \bar{L}(t) \begin{bmatrix} H^T \bar{P} & 0 & \tau H^T R \end{bmatrix} < 0$$

(18)

Apply Lemma 2, a sufficient and necessary condition for (17) to be true is that there exists a positive number λ, such that

$$\begin{bmatrix} \phi_{11} & \phi_{12} & \tau \bar{A}^T R \\ * & X_{22} & \tau(\bar{B}\bar{K})^T R \\ * & * & -\tau R \end{bmatrix} + \lambda \begin{bmatrix} 0 \\ E^T \\ 0 \end{bmatrix} \begin{bmatrix} 0 & E & 0 \end{bmatrix} + \lambda^{-1} \begin{bmatrix} \bar{P}H \\ 0 \\ \tau RH \end{bmatrix} \begin{bmatrix} H^T \bar{P} & 0 & \tau H^T R \end{bmatrix} < 0$$

(19)

Apply Schur complement again, then Equation (18) is equivalent to Equation (12). Therefore, for a given $\tau > 0$ and δ, if there exist $P = P^T > 0$, $Q = Q^T \geq 0$, $R = R^T \geq 0$, $X \geq 0$, and any matrix of appropriate dimensions Y_1 and Y_2, and a scalar $\lambda > 0$, such that for system (9), when LMIS (12) and (13) hold. Thus, there is

$$\dot{J}(t) = \dot{V}(t, y(t)) + \bar{J}_C < 0 \tag{20}$$

According to Lyapunov theory, system (9) is robust and asymptotically stable. In addition, both sides of Equation (19) are integrated from 0 to ∞, and based on the robust asymptotic stability theory, there is

$$J_C = \int_0^\infty \bar{J}_C dt < -V(t,y(t))|_0^\infty < V(y(0)) = y_0^T(I_{N-1} \otimes P)y_0 + \tau y_0^T(Q+R)y_0 = J_C^*. \tag{21}$$

In summary, according to Definition 1, it can be known that Equation (21) is an upper bound of guaranteed cost function (6) in this case.

4. SIMULATIONS

Assume that the form of multi-agent system (1) is as follows:

$$A = \begin{bmatrix} 0 & -0.5 \\ 0.5 & 0 \end{bmatrix}, B = \begin{bmatrix} 0 \\ 1 \end{bmatrix}.$$

The guaranteed cost weighting matrix is defined as follows:

$$Z_x = \begin{bmatrix} 0.6 & 0 \\ 0 & 0.7 \end{bmatrix}, Z_u = I.$$

Consider a system with 10 agents and select the communication delay $d(t) = 0.3(1 - \sin(t))$, which satisfies Assumption 1.

Without loss of generality, assuming that the weight of each edge is 1, the Laplacian matrix of the communication topology between agents is:

$$L = \begin{bmatrix} 4 & 0 & 0 & -1 & -1 & -1 & 0 & 0 & -1 & 0 \\ -1 & 4 & -1 & 0 & 0 & 0 & 0 & 0 & -1 & -1 \\ 0 & -1 & 3 & -1 & 0 & 0 & 0 & -1 & 0 & 0 \\ 0 & 0 & -1 & 3 & 0 & -1 & 0 & 0 & 0 & -1 \\ -1 & 0 & 0 & -1 & 3 & -1 & 0 & 0 & 0 & 0 \\ 0 & 0 & 0 & 0 & -1 & 3 & -1 & 0 & 0 & -1 \\ -1 & 0 & 0 & -1 & 0 & 0 & 3 & -1 & 0 & 0 \\ -1 & 0 & -1 & 0 & 0 & 0 & 0 & 3 & -1 & 0 \\ 0 & -1 & 0 & 0 & 0 & 0 & 0 & -1 & 3 & -1 \\ -1 & 0 & 0 & 0 & 0 & -1 & 0 & 0 & -1 & 3 \end{bmatrix},$$

It can be calculated that the eigenvalues of the matrix $\tilde{S}_0 L \hat{S}_0$ are all negative numbers, indicating that the communication topology is connected, satisfying Assumption 2.

Choosing the uncertain Laplacian matrix $\Delta L = (0.1 \sin t)L$, $L(t) = \sin t$, which satisfies Assumption 3. Without loss of generality, we take the initial state at random $x_0 = [4, 12, 8, 11, 10, 10, 14, 9, 16, 8, 18, 7, 20, 6, 22, 5, 26, 4, 30, 3]^T$, By theorem 3, the gain matrix is calculated $K = \begin{bmatrix} 0 & -0.5 \\ 0.5 & 0 \end{bmatrix}$, which make multi-agent system (1) to achieve guaranteed cost consensus. At this point, the simulation results of the state evolution curve of the multi-agent system (1) are shown in Figure 1.

The state variables $x_{i1}, x_{i2}, i = 1, \cdots, 10$, in Figure 1 represent displacement and velocity of the agent respectively, As can be seen from the simulation results in Figure 1, the uncertain multi-agent system (1) achieves the robust guaranteed cost consensus under the protocol (2). In addition, the consensus convergence sequence of multi-agent system (1) is related to the uncertain parameter ΔL and the communication topology of the system. When the time delays d(t) and ΔL exceed the limited range, the consensus convergence of multi-agent system will be affected.

The guaranteed cost function of the system $J_C \leq J_C^* = 7.21 \times 10^4$ meets the consensus requirement of robust guaranteed cost in Definition 1. Numerical experiments demonstrate that the continuous time high-order multi-agent system (1) can achieve robust guaranteed cost consensus under the protocol (2) with the gain matrix K obtained from Theorem 3, thus verifying the correctness and validity of Theorem 3.

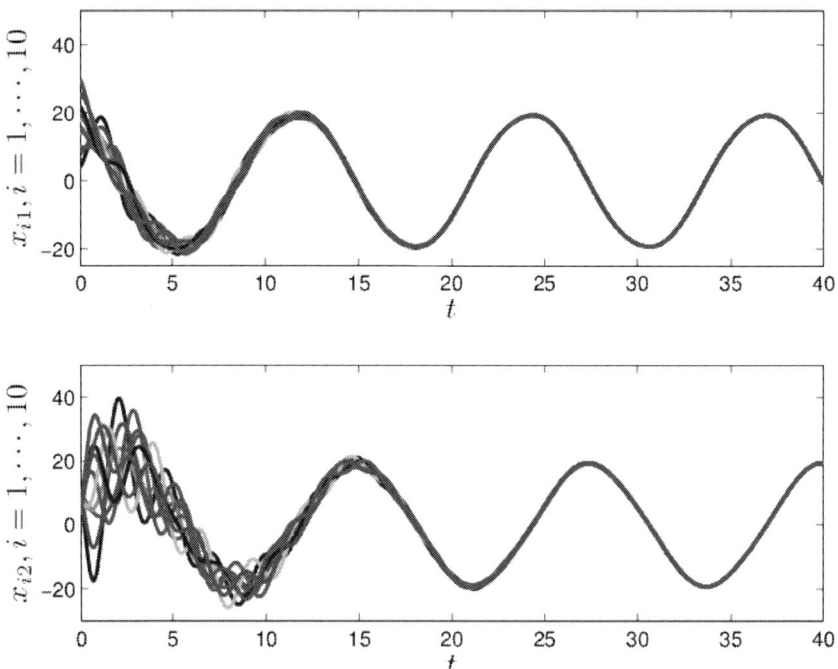

Figure 1. Evolution curve of system (1) to achieve consensus state.

CONCLUSION

In this chapter, based on the directed graph, the continuous-time linear high-order multi-agent systems with time-delay and uncertainty communication topology robust guaranteed cost under the condition of consensus problems were studied, the high-order continuous-time multi-agent system to obtain a sufficient condition for the robust guaranteed cost consensus were given, and the upper bound of the guaranteed cost function were provided. Numerical simulation verifies the correctness of the proposed theory. In the following work, switching communication topology and nonlinear systems will be considered, and the results of this method will be extended to the application field of multi-agent formation control, which will also be the next research goal.

ACKNOWLEDGMENTS

This work is partially supported by Beijing Municipal Science and Technology Project (KM202011417004), Special Research Projects of Beijing Union University (ZK30202002), Science and Technology Program of Beijing Municipal Education Commission (KM201811 417001).

REFERENCES

[1] Chen Y. Z., Ge Y. R., Zhang Y. X. (2014). Partial Stability Approach to Consensus Problem of Linear Multi-Agent Systems. *Acta Automatica Sinica*, 40(3.5), 2573-2584.

[2] Zhang Y. X., Chen Y. Z. (2016). Average dwell-time condition for consensus of linear multi-agent systems with time-varying delay and switching directed communication topologies protocol. *Control and Decision*, 31(2.2), 349-354.

[3] Zhang Y. X. (2016). *Consensus of multiagent with time delays under switching topologies and the applications in formation*. Beijing University of Technology, 47-55.

[4] Cheng Y., Ugrinovskii V. (2013). Guaranteed performance leader-follower control for multi-agent systems with linear IQ Cconstrained coupling. *Proceedings of the 2013 American Control Conference*, 2625-2630.

[5] Wang Z., Xi J. X., Yao Z. C., Liu G. B. (2015). Guaranteed cost consensus for multi-agent systems with flxed topologies. *Asian Journal of Control*, 17(2.2): 729-735.

[6] Wang Z., Xi J. X., Yao Z. C., Liu G. B. (2014). Guaranteed cost consensus problems for second-order multi-agent systems. *Proceedings of the 33rd Chinese Control Conference*, 1069-1074.

[7] Guan Z. H., Hu B., Chi M., He D. X., Cheng X. M. (2014). Guaranteed performance consensus in second-order multi-agent systems with hybrid impulsive control. *Automatica*, 50(9): 2415-2418.

[8] Xu J., Zhang G. L., Zeng J., Tang W. J., Huang X. (2016). Guaranteed cost consensus analysis of discrete-time high-order uncertain linear multi-agent systems. *Control Theory and Application*, 33(2.6): 841-848.

[9] Xu J., Zhang G. L., Zeng J., Sun Q., (2019). Yang. Robust guaranteed cost consensus for high-order discrete-time multi-agent systems with switching topologies and time delays. *Acta Automatica Sinica*, 45(02):129-142.

[10] Liu S. (2008). *Schur Complement, Encyclopedia of Statistical Sciences.*

[11] Petersen I. R., Hollot C. V. (1986). A riccati equation approach to the stabilization of uncertain linear systems. *Automatica*, 22(4):397-411.

In: Networked Control Systems
Editors: S. Tong and D. Qian
ISBN: 978-1-53619-892-8
© 2021 Nova Science Publishers, Inc.

Chapter 7

ROBUST GUARANTEED PERFORMANCE FORMATION CONTROL FOR MAS WITH UNCERTAIN TOPOLOGIES

Yaxiao Zhang[1,*] and Shiwen Tong[1,2]

[1]College of Robotics, Beijing Union University, Beijing, China
[2]State Key Laboratory for Management and Control of Complex Systems, Institute of Automation, Chinese Academy of Sciences, Beijing, China

ABSTRACT

This chapter investigates a robust guaranteed performance formation problem for a class of continuous-time linear high-order multi-agent systems with uncertain communication topology which is modeled by directed graph. Firstly, the robust guaranteed performance formation problem is transformed into a robust guaranteed performance control problem of an auxiliary uncertain system by a linear transformation. Secondly, a sufficient condition for the robust guaranteed performance formation control problem is presented in terms of linear matrix inequality techniques, and an upper bound of the guaranteed performance function is given. Finally, a numerical example is shown to demonstrate the effectiveness of the theoretical results.

---

* Corresponding Author's E-mail: zdhtyaxiao@buu.edu.cn.

Keywords: robust guaranteed performance, multi-agent systems, formation, uncertain topologies

1. INTRODUCTION

The distributed cooperative control of multi-agent systems has been studied extensively due to its wide applications in different fields such as formation [1-3], synchronization [4, 5], flocking [6, 7]. Motivated by the formation flying phenomenon in the biological world, formation control has already been a hotspot of research in both academic and engineering community in the past two decades.

Many researchers have considered formation control problems for a group of agents and many traditional methods were proposed to analyze the formation control for multi-agent systems [8-14]. However, in practical applications, multi-agent systems are not only required to achieve some formation performance but also may have limited energy supply to complete their task, such as communication, sensing, and movement. Thus, it is a key technique to realize a trade off between formation performance and energy consumption, which can usually be modeled as optimal or suboptimal formation problems.

To the best of our knowledge, there are few papers addressing guaranteed performance formation problems for multi-agent systems. Guaranteed performance control was first introduced into formation control problems for multi-agent systems with undirected graph in [15], the guaranteed performance formation control problem was transformed to the guaranteed performance consensus problem by the using consensus algorithm, and a sufficient condition for the guaranteed performance formation control problem as well as an upper bound of the guaranteed performance function were proposed. Guaranteed performance time-varying formation analysis and design problems for general high-order multi-agent systems with communication constraints under undirected graph were investigated in [16], a guaranteed performance time-varying formation control protocol was proposed by using the intermittent relative local information, sufficient conditions for guaranteed-performance time-varying formation analysis and design were respectively given as well as a guaranteed-performance cost function.

However, in many practical cases, communication uncertainties in information transmission may occur due to network-induced packet loss, external disturbances, or faults of sensing devices. Thus, robustness is very important for multi-agent systems. The robust consensus problem was converted to the robust stability analysis of the error model based on robust Lyapunov stability theory in [17-19]. In [20], it was shown that a robust formation control problem of a class of multi-agent systems based on the generalized internal model principle is boils down to a robust stabilization problem of a nonlinear system, and proposed a design method for one typical formation control problem of multi-agent systems. A robust formation control problem of discrete-time multi-agent systems with unknown nonlinear dynamics was considered in [21], and the formation problem is studied by converting into a stability control problem.

Motivated by this, the robust guaranteed performance formation control problem of high-order continuous-time linear multi-agent systems with uncertain communication topology is investigated in this chapter. Compared with [17-21], guaranteed cost control is introduced into formation control problems for multi-agent systems, where the formation performance and the control energy consumption are considered simultaneously. Compared with [15, 16], the main contribution of the current chapter is that the uncertain topologies under directed graphs is considered. The core idea of this chapter is to transform the robust formation control problems into robust stability problems equivalently by a novel linear transformation, and then solve the problem by the tools in robust stability theory in terms of linear matrix inequalities (LMIs), moreover an upper bound of the guaranteed performance function is formulated.

The rest of the chapter is organized as follows. Section 2 shows the problem description based on graph theory. Sections 3 presents a linear transformation approach to transform the robust formation control problem into a robust stability problem of a corresponding auxiliary system, Section 4 derives a sufficient condition for robust guaranteed-cost formation, and the upper bound of guaranteed performance are presented. Numerical results are presented in Section 5, and then concluding remarks are given.

Notations: $\mathbb{R}^n$ and $\mathbb{R}^{n\times m}$ are the n-dimension real column vector and the set of $n \times m$ dimensional real matrices, respectively. Let 0 be zero number, zero vectors, or zero matrices in appropriate dimension, respectively. Let 1_N denote an N-dimensional column vector with 1. Let P^T and P^{-1} denote the transpose and the inverse matrix of P, respectively. $P^T = P > 0$ stands for matrix P is symmetric and positive definite. The notation * denotes the symmetric terms of a symmetric matrix. $\otimes$ is the Kronecker product of matrices.

2. PROBLEM STATEMENT

Consider a system consisting of N agents, where each agent takes the following dynamics:

$$\dot{x}_i(t) = Ax_i(t) + Bu_i(t), \; x_i(0) = x_{i0}, \; i \in \mathrm{I} \tag{1}$$

with

$$A = \begin{bmatrix} 0 & I_n \\ 0 & 0 \end{bmatrix} \in \mathbb{R}^{2n \times 2n}, B = \begin{bmatrix} 0 \\ I_n \end{bmatrix} \in \mathbb{R}^{2n \times m},$$

where $x_i(t) = [s_i^T(t), v_i^T(t)]^T$, $s_i(t) \in \mathbb{R}^n$ and $v_i(t) \in \mathbb{R}^n$ are the position state and the velocity state of agent i respectively, $u_i(t)$ is the control protocol of agent i, which depends on x_i and x_j, agent j is called a neighbor of agent i if there exists a communication channel from j to i, and $\mathrm{I} = \{1, \cdots, N\}$ is the index set of agents.

Denote $N_i(t)$ be the set of the neighbors of the agent i at time t, and $N(t) = \{N_i(t), i = 1, \cdots, N\}$ a communication configuration of the system (1) at time t. $N(t)$ can be expressed by a digraph $G = (V, E(t), W(t))$. Vertex set $V = \{1, 2, \cdots, N\}$ represents the group of agents, time-varying edge set $E(t) \subseteq V \times V$ denotes the communication topology $N(t)$, i.e. $(j,i) \in E(t) \leftrightarrow j \in N_i(t)$, and $W(t) = [w_{ij}] \in \mathbb{R}^{N \times N}$ is a weighted adjacency matrix.

A formation, which described by a vector $H = [h_1^T, h_2^T, \cdots, h_N^T]^T \in \mathbb{R}^{2nN}$, is a geometric pattern, it satisfies some predefined geometric constraints which is required to achieve and maintain for the LMAS (1). H represents the desired formation.

For a desired formation H, a formation protocol is considered as follows:

$$u_i(t) = K \sum_{j \in N_i(t)} (\omega_{ij} - \Delta\omega_{ij}(t))[(x_j(t) - h_j) - (x_i(t) - h_i)], t > 0, \qquad (2)$$

where $\omega_{ij}(t) = \omega_{ij} - \Delta\omega_{ij}(t)$ represents the coupling strength with respect to a communication channel from j to i at time t, $\Delta\omega_{ij}(t)$ represents parameter uncertainties in the communication channel.

Definition 1 Linear multi-agent system (LMAS) (1) is said to achieve formation H robustly, if there exist vector valued functions $\xi(t) \in \mathbb{R}^{2n}$ and a control protocol (2), such that $\lim_{t \to \infty} \|x_i(t) - h_i\| = \xi(t), i \in I$, and the vector valued function $\xi(t) = [\xi_s^T(t) \quad \xi_V^T(t)]^T$ is called a formation center function. Remark 1: In the current chapter, the formation vector $h_i = [h_{is}^T \quad h_{iv}^T]^T$ is used to express the relative position h_{is} and the relative velocity h_{is} of agent respectively. It is generally known that the velocity state $v_i(t)$ are synchronous when multi-agent system (1) achieves the formation H, hence, there is $H = [h_1^T, h_2^T, \cdots h_N^T]^T$ with $h_i = [h_{si}^T] \in \mathbb{R}^{2n}, i = 1, \cdots, N$. Consider the following linear quadratic cost function

$$J_C = \sum \int_0^\infty \{\sum w_{ij}(t)[(x_j(t) - h_j) - (x_i(t) - h_i)]^T Q \\ \cdot [(x_j(t) - h_j) - (x_i(t) - h_i)] + u_i^T(t) R u_i(t)\} dt, \qquad (3)$$

where Q and R are given symmetric positive matrices.

Definition 2 LMAS (1) is said to achieve robust guaranteed performance formation H under the protocol (2), if for any initial condition sequence $x(0)$ there is $\lim\limits_{t\to\infty}\|(x_i(t)-h_i)-(x_j(t)-h_j)\|=0, i,j \in I$ and there exists a $J_C^* > 0$ such that $J_C \leq J_C^*$, J_C^* is said to be a guaranteed performance.

Definition 3 LMAS (1) with respect to the formation H is said to be robust guaranteed performance feasible under formation protocol (2), if there exist control gain matrix K such that multi-agent system (1) achieves robust guaranteed performance formation H.

Let $x = [x_1^T \cdots x_N^T]^T \in \mathbb{R}^{2nN}$, and the dynamics of the LMAS (1) with formation protocol (2) can be described by a compact form as follows:

$$\dot{x}_i(t) = [I_N \otimes A]x(t) - [L - \Delta L(t) \otimes BK](x(t) - H), \qquad (4)$$

where Outer-coupling matrix $L = [l_{ij}] \in \mathbb{R}^{N \times N}$ is Laplacian matrix induced by the communication topology, $N(t) = \{N_i(t), i = 1, \cdots, N\}$ and its entries are defined by

$$l_{ij} = \begin{cases} \sum\limits_{k \in N_i} \omega_{ik}, & j = i \\ -\omega_{ij}, & j \neq i, j \in N_i \\ 0, & j \notin N_i \end{cases}$$

The communication topology N and the Laplacian matrix $\Delta L(t)$ satisfy the following condition:

Assumption 1 There exist a directed spanning tree for dynamic digraph G.

Assumption 2 Uncertain Laplacian matrix $\Delta L(t)$ can be expressed as follows

$$\Delta L(t) = L^1 L(t) L^2, \qquad (5)$$

where L^1 and L^2 are constant matrices of appropriate dimension, $L(t)$ is an unknown time-varying matrix which satisfies the following condition:

$$\mathbf{L}^T(t)\mathbf{L}(t) \le I. \tag{6}$$

Next, we equivalently transform the robust formation problem of a multi-agent system with uncertain communication topologies into a robust stability problem of a corresponding auxiliary uncertain systems by a linear transformation, and analyze the sufficient condition that system (1) can achieve formation H via protocol (2) in terms of robust stability theory.

3. A LINEAR TRANSFORMATION

Transform system (4) by the following linear transformation:

$$\bar{x}(t) = S(x(t) - H), \tag{7}$$

where

$$S = \begin{bmatrix} \tilde{S}_0 \\ 1_N^T \end{bmatrix} \otimes I_{2n}, \quad \tilde{S}_0 = \begin{bmatrix} 1 & -1 & 0 & \cdots & 0 \\ 0 & 1 & -1 & \cdots & 0 \\ \vdots & \ddots & \ddots & \ddots & \vdots \\ 0 & \cdots & 0 & 1 & -1 \end{bmatrix}.$$

The inverse matrix of S can be worked out as follows:

$$S^{-1} = \frac{1}{N} \begin{bmatrix} N-1 & N-2 & \cdots & 1 & 1 \\ -1 & N-2 & \cdots & 1 & 1 \\ \vdots & \vdots & \ddots & \vdots & \vdots \\ -1 & -2 & \cdots & 1 & 1 \\ -1 & -2 & \cdots & -(N-1) & 1 \end{bmatrix} \otimes I_{2n} = \begin{bmatrix} \hat{S}_0 & N^{-1}1_N \end{bmatrix} \otimes I_{2n}.$$

By the linear transformation (7), system (4) is transformed into the following system:

$$\dot{\bar{x}}(t) = S[(I_N \otimes A) - (L - \Delta L(t)) \otimes BK]S^{-1}\bar{x}(t) + S(I_N \otimes A)H, \quad (8)$$

Let $\bar{x} = \begin{bmatrix} y^T & z^T \end{bmatrix}^T$, where $y = [\bar{x}_1^T \cdots \bar{x}_{N-1}^T]^T$, $z = \bar{x}_N$, then, the following lemma presents that the formation problem of multi-agent system can be transformed to the robust stability problem of a lower dimension system. Unfold system (8) by $\bar{x} = \begin{bmatrix} y^T & z^T \end{bmatrix}^T$:

$$\dot{\bar{x}} = \begin{bmatrix} \dot{y}(t) \\ \dot{z}(t) \end{bmatrix}$$

$$= (\begin{bmatrix} \tilde{S}_0 \\ 1_N^T \end{bmatrix} \otimes I_{2n})[(I_N \otimes A) - ((L - \Delta L(t) \otimes BK)](\begin{bmatrix} \hat{S}_0 & N^{-1}1_N \end{bmatrix} \otimes I_{2n}) \begin{bmatrix} y(t) \\ z(t) \end{bmatrix} + \begin{bmatrix} (\tilde{S}_0 \otimes A)H \\ (1_N^T \otimes A)H \end{bmatrix}$$

$$= \begin{bmatrix} \wp_{11} & 0 \\ \wp_{21} & A \end{bmatrix} \begin{bmatrix} y(t) \\ z(t) \end{bmatrix} + \begin{bmatrix} (\tilde{S}_0 \otimes A)H \\ (1_N^T \otimes A)H \end{bmatrix},$$

where

$$\wp_{11} = I_{N-1} \otimes A - (\tilde{S}_0 (L - \Delta L(t)\hat{S}_0) \otimes BK,$$

$$\wp_{21} = -1_N^T (L - \Delta L(t)\hat{S}_0) \otimes BK.$$

System (8) is equivalent to the following system:

$$\begin{cases} \dot{y}(t) = [I_{N-1} \otimes A - (\tilde{S}_0 (L - \Delta L(t)\hat{S}_0) \otimes BK]y(t) + (\tilde{S}_0 \otimes A)H \\ \dot{z}(t) = Az(t) - 1_N^T (L - \Delta L(t)\hat{S}_0) \otimes BKy(t) + (1_N^T \otimes A)H \end{cases}$$

Let $\bar{A} = I_{N-1} \otimes A$, $\bar{B} = -(\tilde{S}_0 L \hat{S}_0) \otimes B$, $\Delta \bar{B} = (\tilde{S}_0 \Delta L(t)\hat{S}_0) \otimes B$, $\bar{K} = I_{N-1} \otimes K$,

Rewrite $\Delta \bar{B} = H\bar{L}(t)E$, where $\bar{L}^T(t)\bar{L}(t) \leq I$.

Then the above equation can be expressed by

$$\begin{cases} \dot{y}(t) = (\bar{A} + \bar{B}\bar{K} + H\bar{L}(t)E\bar{K})y(t) + (\tilde{S}_0 \otimes A)H \\ \dot{z}(t) = Az(t) - 1_N^T \left(L - \Delta L(t)\hat{S}_0\right) \otimes BKy(t) + (1_N^T \otimes A)H \end{cases} \quad (9)$$

The following Lemma transform the formation problem to a robust stability problem equivalently.

Lemma 1 LMAS (1) achieves formation H robustly with protocol (2) for any bounded initial states $x(0)$ if and only if the system

$$\dot{y}(t) = (\bar{A} + \bar{B}\bar{K} + H\bar{L}(t)E\bar{K})y(t) \quad (10)$$

is robustly stable.

Proof: From Definition 1, one can see that LMAS (1) achieves formation robustly with protocol (2) if there exist vector-valued function $\xi(t)$ such that

$$\lim_{t \to \infty} \| x_i(t) - h_i \| = \xi(t), i \in \mathcal{I}$$

that is

$$\lim_{t \to \infty} \| (x_i(t) - h_i) - (x_j(t) - h_j) \| = 0, i, j \in \mathcal{I}$$

it is equivalent to

$$\lim_{t \to \infty} \| (\tilde{S}_0 \otimes I_{2n})(x(t) - H) \| = 0,$$

that is $\lim_{t \to \infty} \| y(t) \| = 0$ of system (10).

Hence, LMAS (1) achieves formation robustly with protocol (2) is equivalent to $\lim_{t \to \infty} \| y(t) \| = 0$ of system (10).

According to the structure of A and H, it derived that

$(\tilde{S}_0 \otimes A)H = 0$

in the first equation of system (9).

Moreover, there is no relationship between $y(t)$ and $z(t)$ in the first equation in system (9).

Therefore, LMAS (1) achieves formation H with protocol (2) robustly for any bounded initial states $x(0)$ if and only if $\lim_{t \to \infty} \|y(t)\| = 0$ of system (10).

Hence, LMAS (1) achieves formation H with protocol (2) robustly for any bounded initial states $x(0)$ if and only if system (10) is robustly stable.

Cost function (3) can be rewritten as follows:

$$J_C = \int_0^\infty y^T(t)\{2(L-\Delta L) \otimes Q + [(L-\Delta L)^T(L-\Delta L)] \otimes (K^T RK)\}y(t)dt. \quad (11)$$

According to Lemma 1 and Definition 2, the following theorem can be obtained:

Theorem 1 LMAS (1) achieves robust guaranteed performance formation H with protocol (2) for any bounded initial states $x(0)$, if and only if (10) is robustly stable and there exists a $J_C^* > 0$, such that $J_C < J_C^*$.

4. MAIN RESULTS

Lemma 3 [22] (Schur Complement) For given symmetric matrix $P \in \mathbb{R}^{(m+n) \times (m+n)}$:

$$P = P^T = \begin{bmatrix} P_{11} & P_{12} \\ * & P_{22} \end{bmatrix},$$

where $P_{11} \in \mathbb{R}^{m \times m}$, $P_{22} \in \mathbb{R}^{n \times n}$. Then the following three conditions are equivalent:

(2.1) $P < 0$;

(2.2) $P_{11} < 0, P_{22} - P_{12}^T P_{11}^{-1} P_{12} < 0$;

(2.3) $P_{22} < 0, P_{11} - P_{12}^T P_{11}^{-1} P_{12}^T < 0$.

Lemma 4 [23] For given matrices $Q = Q^T, H, E$, there is

$$Q + HL(t)E + E^T L^T(t)H^T < 0.$$

For any $L(t)$ that satisfies $L^T(t)L(t) \le I$ if and only if there exist a $\varepsilon > 0$ such that

$$Q + \varepsilon^{-1}HH^T + \varepsilon E^T E < 0.$$

Theorem 2 Suppose Assumption 1 and Assumption 2 hold, LMAS (1) with respect to the formation H is robust guaranteed performance feasible by formation protocol (2) if there exists a $2n(N-1)$-dimensions matrix $X = X^T > 0$ and matrix W such that

$$\Phi = \begin{bmatrix} \phi_{11} & E\bar{K} & I & L \otimes K \\ * & -\varepsilon I & 0 & 0 \\ * & * & -2L \otimes Q^{-1} & 0 \\ * & * & 0 & -R^{-1} \end{bmatrix} < 0, \quad (12)$$

where

$$\phi_{11} = \bar{A}X + \bar{B}W + (\bar{A}X + \bar{B}W)^T + \varepsilon HH^T.$$

In this case, the control gain matrix in formation protocol (2) satisfies $K = WX^{-1}$, and guaranteed performance $J_C^* = y_0^T X^{-1} y_0$

Proof: According to Theorem 1, we only need to prove $\lim_{t\to\infty} y(t)=0$ in system (10). Consider the following Lyapunov function candidate of system (10)

$$V(y(t)) = y^T(t)Py(t), \qquad (13)$$

where $2n(N-1)$-dimension matrix $P = P^T > 0$, it follows that $V(y(t)) \geq 0$ and $V(y(t))=0$ if and only if $y(t)=0$. Then the time derivative of $V(t)=V(y(t))$ with respect to t along the trajectory of system (10) is

$$\dot{V}(t) = y^T(t)[P(\bar{A}+\bar{B}\bar{K}+H\bar{L}(t)E)+(\bar{A}+\bar{B}\bar{K}+H\bar{L}(t)E)^T P]y(t) \qquad (14)$$

Next, define

$$\dot{\Im}(t) = \dot{V}(t) + \bar{J}_C, \qquad (15)$$

where $\bar{J}_C \geq 0$ and

$$\bar{J}_C = y^T(t)\{2(L-\Delta L)\otimes Q + [(L-\Delta L)^T(L-\Delta L)]\otimes (K^T RK)\}y(t). \qquad (16)$$

Hence, there is

$$\begin{aligned}\dot{\Im}(t) &= \dot{V}(t) + \bar{J}_C \\ &= y^T(t)[P(\bar{A}+\bar{B}\bar{K}+H\bar{L}(t)E)+(\bar{A}+\bar{B}\bar{K}+H\bar{L}(t)E)^T P] \\ &\quad + 2(L-\Delta L)\otimes Q + [(L-\Delta L)^T(L-\Delta L)]\otimes (K^T RK)\}y(t) \\ &= y^T(t)Yy(t).\end{aligned} \qquad (17)$$

Let

$$Y = 2(L-\Delta L)\otimes Q + [(L-\Delta L)^T(L-\Delta L)]\otimes (K^T RK) + P(\bar{A}+\bar{B}\bar{K}) + (\bar{A}+\bar{B}\bar{K})^T P,$$

it has

$$\eta = Y + PH\bar{L}(t)E\bar{K} + (E\bar{K})^T \bar{L}^T(t)(PH)^T$$
$$-2\Delta L \otimes Q - (L^T \Delta L + \Delta L^T L - \Delta L^T \Delta L] \otimes (K^T RK)$$
$$< Y + PH\bar{L}(t)E\bar{K} + (E\bar{K})^T \bar{L}^T(t)(PH)^T = \tilde{Y}.$$

If $\tilde{Y} < 0$ according to lemma 4, there is

$$Y + \varepsilon PHH^T P + \varepsilon^{-1}(E\bar{K})^T(E\bar{K}) < 0. \tag{18}$$

From Lemma 3, inequation (17) is further equivalent to

$$\tilde{\Phi} = \begin{bmatrix} \tilde{\phi}_{11} & E\bar{K} & I & L \otimes K \\ * & -\varepsilon I & 0 & 0 \\ * & * & -2L \otimes Q^{-1} & 0 \\ * & * & 0 & -R^{-1} \end{bmatrix}, \tag{19}$$

where

$$\tilde{\phi}_{11} = P(\bar{A} + \bar{B}\bar{K}) + (\bar{A} + \bar{B}\bar{K})^T P + \varepsilon PHH^T P.$$

Multiple matrix $diag\{P^{-1}, I, I, I\}$ from left side and right side for inequation (19), let $X = P^{-1}$, $W = \bar{K}P^{-1}$, we can get inequation (12). Therefore, if there exist a $2n(N-1)$-dimensions matrix $X = X^T > 0$ and matrix w such that inequation (12) is true, it has

$$Y < \tilde{Y} < 0,$$

that is

$$\tilde{\mathfrak{J}}(t) = \dot{V}(t) + \bar{J}_C = y^T(t)Yy(t) < 0.$$

For all the permitted uncertainty, there is

$$\dot{V}(T) < -\bar{J}_C < 0 \tag{20}$$

From Lyapunov theory, system (10) is robust asymptotically stable. Moreover, integrating both side of (20) over time t from 0 to ∞, and on the basis of robust asymptotically stability, there is

$$J_C = \int_0^\infty \bar{J}_C dt < -V(y(t))\big|_0^\infty < V(y(0)) = y_0^T P y_0 = y_0^T X^{-1} y_0 = J_C^*.$$

Therefore, LMAS (1) with respect to the formation H is robust guaranteed performance feasible by formation protocol (2), and the guaranteed performance $J_C^* = y_0^T X^{-1} y_0$ The conclusion of theorem 2 can be obtained.

5. SIMULATIONS

In this section, a numerical example is given to illustrate the effectiveness of the obtained theoretical results. We apply the above proposed formation protocol to achieve state alignment among 4 agents. The dynamics of them are described by

$$\dot{x}_i(t) = \begin{bmatrix} 0 & 1 \\ 0 & 0 \end{bmatrix} x_i(t) + \begin{bmatrix} 0 \\ 1 \end{bmatrix} u_i(t), i = 1, \cdots, 4, \tag{21}$$

where $x_i(t) = [x_{i1}(t) \quad x_{i2}(t)]^T$, $x_{i1}(t)$ and $x_{i2}(t)$ denote position and velocity of agent i respectively.

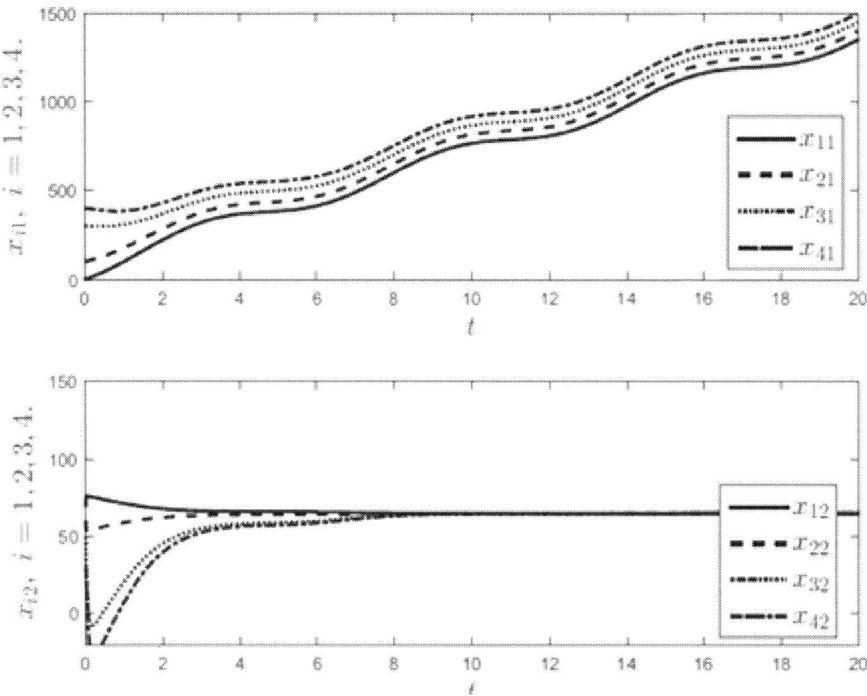

Figure 1. Position and velocity state trajectories of system (5.1).

The initial state is chosen randomly:

$$x(0) = \begin{bmatrix} 0 & 70 & 100 & 60 & 300 & 50 & 400 & 30 \end{bmatrix}^T.$$

Without loss of generality, let the communication topology weight is 1, assume the communication topology is N with Laplacian matrix as follows:

$$L = \begin{bmatrix} 1 & -1 & 0 & 0 \\ -1 & 1 & 0 & 0 \\ 0 & -1 & 2 & -1 \\ 0 & 0 & -1 & 1 \end{bmatrix},$$

which satisfies Assumption 1. We choose the uncertain topology $\Delta L = (\alpha \sin t)L$, hence $L(t) = \sin t$ satisfies Assumption 2, choosing $\alpha = 0.8$.

In the guaranteed-performance cost function (3) we choose $Q = 0.6I_2$ and $Q = 0.4I_2$. The target formation $H = [50 \ 100 \ 150 \ 200]^T \otimes [1 \ 0]^T$. It can be verified that $(\tilde{S}_0 \otimes A)H = 0$ in system (9).

From Theorem 2, it can be obtained that the gain matrix $K = [18.312 \ 42.517]$ and $J^* = 8.791 \times 10^5$ are solvable by LMIs. Figure 1 shows the state trajectories of system (21) via protocol (2) with uncertain topology $\Delta L = (0.8 \sin t)L$. From Figure 1 we can see that LMAS (21) can achieve robust guaranteed performance formation H with protocol (2).

CONCLUSION

The robust guaranteed performance formation control for multi-agent systems with uncertain topologies under directed graphs was investigated in the current chapter. Robust guaranteed performance control for multi-agent systems were transformed into robust stability problem of an auxiliary system. A sufficient condition for the robust guaranteed performance formation control was given and an upper bound of the guaranteed performance function was determined. Finally, the effectiveness of the proposed theory has been illustrated by a 4 agents system experiments. Further research will be conducted to the guaranteed performance formation problem of multi-agent systems with switching topologies and time-delays.

ACKNOWLEDGMENTS

This work is partially supported by Beijing Municipal Science and Technology Project (KM202011417004), Special Research Projects of Beijing Union University (ZK30202002), Science and Technology

Program of Beijing Municipal Education Commission (KM201811417001).

REFERENCES

[1] Cai N., He M., Wu Q. X., Khan M. J. (2019). On Almost controllability of dynamical complex networks with noises. *Journal of SystemsScience & Complexity*, 32:1125-1139.

[2] Dong X., Zhou Y., Ren Z., et al., (2017). Time-varying formation tracking for second-order multi-agent systems subjected to switching topologies with application to quadrotor formation flying. *IEEE Transactions on Industrial Electronics*, 64(2.6):5014-5024.

[3] Xue R., Cai G. (2016). Formation flight control of multi-UAV system with communication constraints. *J. Aerosp. Technol. Manag.* 8(2.2):203-210.

[4] Xi J., Wang C., Liu H., and Wang Z. (2018). Dynamic output feedback guaranteed-cost synchronization for multiagent networks with given cost budgets. *IEEE Access*,6(2.1):28923-28935.

[5] Almeida J., Silvestre C., and Pascoal A. (2017). Synchronization of multi-agent systems using event-triggered and self-triggered broadcasts. *IEEE Transactions on Automatic Control*, 62(3.3):4741-4746.

[6] Cai N., Diao C., Khan M. J. (2017). A novel clustering method based on quasi-consensus motions of dynamical multi-agent systems. *Complexity*, 1-8.

[7] Zhou J., Wu X., Yu W., et al., (2012). Flocking of multi-agent dynamical systems based on pseudo-leader mechanism. *Systems & control letters,* 61(2.1):195-202.

[8] Lafferriere G., Williams A., Caughman J., Veerman J. J. P. (2005). Decentralized control of vehicle formations. *Systems & controlletters*, 54(3.3): 899-910.

[9] Ren W., and Sorensen N. (2008). Distributed coordination architecture for multi-robot formation control. *Robotics and Autonomous Systems*, 56(2.4): 324-333.

[10] Xie G., and Wang L. (2009). Moving formation convergence of a group of mobile robots via decentralised information feedback. *International Journal of Systems Science*, 40(3.4): 1019-1027.

[11] Ma C., and Zhang J. (2012). On formability of linear continuous-time multi-agent systems. *Journal of Systems Science and Complexity*, 25(2.1): 13-29.

[12] Dong X. W., Xi J. X., et al., (2014). Formation control for high-order linear time-invariant multi-agent systems with time delays. *IEEE Transactions on Control of Network Systems*, 1(2.3): 232-240.

[13] Yu B., Dong X. W., Shi Z., et al., (2013). Formation control for quadrotor swarm systems: Algorithms and experiments. *Proceedings of Chinese Control Conference*, 7099-7104.

[14] Zhang Y. X., Chen Y. Z., Qu X. J. (2016). Design of topology switching law for formation problem of linear multi-agent systems with timevarying delay. *Proceedings of Chinese Control Conference*, 8002-8007.

[15] Wang Z., Liu G., Xi J., et al., (2015). Guaranteed cost formation control for multi-agent systems: Consensus approach. *Proceedings of the 34th Chinese Control Conference,* 7309-7314.

[16] Wang L., Xi J., Yuan M., Liu G. (2018). Guaranteed-performance time-varying formation control for swarm systems subjected to communication constraints. *IEEE Access*, 6:45384-45393.

[17] Huang W. C., Zeng J. P., Sun H. F. (2015). Robust consensus for linear multi-agent systems with mixed uncertainties. *Systems & control letters* 76:56-65.

[18] Li Z. K., Chen J. (2017) Robust consensus of linear feedback protocols over uncertain network graphs. *IEEE Transactions on Automatic Control* 62(3.2): 4251-425.

[19] Huang W., Huang Y., Chen S. (2018). Robust consensus control fora class of second-order multi-agent systems with uncertain topology and disturbances. *Neurocomputing*, 31:3426-435.

[20] Sun W., Qiao Y., Suo X. (2012). Robust formation control of a class of multi-agent systems by output regulation approach. *Proceedings of the IEEE Intelligent Control & Automation.*

[21] Liu Y., Jia Y. (2015). Robust formation control of discrete-time multiagent systems by iterative learning approach. *International Journal of Systems Science*, 46(2.4): 625-633.

[22] Liu S. (2008). *Schur Complement,* Encyclopedia of Statistical Sciences.

[23] Petersen I. R., Hollot C. V. (1986). A riccati equation approach to the stabilization of uncertain linear systems. *Automatica*, 22(2.4):397-411.

ABOUT THE EDITORS

Shiwen Tong
Professor
College of Robotics,
Beijing Union University, Beijing China
Email: shiwen.tong@buu.edu.cn

 Shiwen Tong received the B.E. degree in chemical engineering from the University of Petroleum (East China), Shandong, China, in 1999, the M.E. degree in control theory and control engineering from the University of Petroleum (Beijing), Beijing, China, in 2003, and the Ph.D. degree from the Institute of Automation, Chinese Academy of Sciences, Beijing, China, in 2008. He was an operator with Liaohe Oil Feild Petrochemical Refinery from 1999–2002, an Engineer with Bejjing Anwenyou Science and Technology Company, Ltd. from 2003–2005, and an instrument Senior Engineer with China Tianchen Engineering Corporation (TCC) from 2008–2012.He is currently a Professor with the College of Robotics, Beijing Union University, Beijing, China. He has authored more than forty academic papers in IEEE Transactions on Control Systems Technology, European Journal of Control, and Neurocomputing etc. He has also published five monographs, two book chapters and seven patents. His research interests include the intelligent control, networked control, PEM fuel cell, and their industrial applications.

Dianwei Qian
Associate Professor
School of Control and Computer Engineering,
North China Electric Power University, Beijing, China
Email: dianwei.qian@ncepu.edu.cn

Dianwei Qian received his BS degree form Hohai University, Nanjing, China, in 2003. He received his MS degree from Northeastern University, Shenyang, China, and his PhD from the Institute of Automation, Chinese Academy of Sciences, Beijing, China, in 2005 and 2008, respectively. He is currently an associate professor at the School of Control and Computer Engineering, North China Electric Power University, Beijing, China. His research interests include the theory and applications of intelligent and nonlinear control. He has authored or coauthored over 80 papers in international journals and conferences. He has been a program committee member of several international conferences, and a reviewer for several international conferences and journals.

INDEX

A

absolute triggering condition, 80, 101
A_i, 23, 24, 34, 45, 66
air tank, v, ix, 59, 60, 61, 68, 69, 71
algorithm, ix, 7, 10, 15, 21, 23, 24, 30, 33, 42, 44, 45, 48, 50, 51, 52, 55, 59, 64, 68, 69, 71, 134

B

b_i, 23, 25, 34, 45, 47, 48, 49

C

chattering Zeno, 81, 83, 84, 85, 86, 89, 90, 91, 94, 95, 96
cluster model, 20, 23, 30, 44, 47
clustering, ix, 23, 39, 42, 43, 45, 46, 47, 56, 149
communication, viii, ix, x, 2, 3, 12, 15, 17, 23, 28, 31, 34, 38, 42, 50, 52, 73, 74, 79, 116, 117, 118, 119, 120, 127, 128, 129, 130, 131, 133, 134, 135, 136, 137, 138, 139, 147, 149, 150
communication delay, 12, 116, 117, 118, 120, 127, 128
compensate, 3, 4, 12, 37, 49, 50, 54
compensation, vii, ix, 2, 4, 13, 15, 20, 21, 40, 42, 43, 44, 57
configuration, ix, 43, 44, 59, 71, 118, 136
consensus, vi, x, 111, 113, 114, 115, 116, 117, 118, 120, 121, 122, 123, 129, 130, 131, 132, 134, 135, 149, 150
control performance, vii, viii, 1, 2, 8, 10, 13, 15, 16, 21, 31, 32, 34, 35, 42, 52, 53, 54, 63, 66, 69, 71, 74
convergence, 88, 101, 103, 104, 110, 129, 150
cost, 2, 23, 117, 119, 120, 123, 128, 129, 130, 131, 132, 135, 137, 148, 149, 150

D

data-based, v, vii, viii, 19, 20, 21, 30, 31, 39, 41, 43, 54
decomposition, 116
delay compensation, vii, 2, 4, 13, 15, 20, 28, 42, 50, 57

digraph, 118, 136, 138
diophantine equation, 3, 5, 6, 16
directed spanning tree, 138
distributed coordinated control, 116
DNTC, vii, 21, 22, 31, 32, 34, 35, 37
dynamical systems, 77, 112, 149

E

energy, 74, 116, 134, 135
energy consumption, 116, 134, 135
energy supply, 116, 134
engineering, 71, 134, 153
equilibrium, 75, 82, 83, 84, 86, 87, 88, 89, 90, 91, 92, 93, 94, 95, 101, 102, 103, 104, 106, 110, 121
equipment, 69
event-triggered control, viii, ix, 73, 74, 75, 76, 80, 81, 82, 83, 84, 86, 93, 94, 95, 96, 97, 98, 99, 100, 101, 102, 103, 107, 109, 110, 111, 112, 113, 114
excitation, viii, 30, 33, 41, 43, 44

F

fault detection, 55
finite-time stability, ix, 74, 76, 100, 101, 103, 110, 114
formation, vi, x, 116, 131, 133, 134, 135, 137, 138, 139, 140, 141, 142, 143, 146, 148, 149, 150, 151
formation protocol, 137, 138, 143, 146
fuzzy cluster modeling, 20, 30, 44
fuzzy clustering, ix, 23, 39, 42, 43, 46, 47, 56
fuzzy control, viii, 2, 3, 4, 7, 9, 10, 11, 13, 15, 16, 17, 18, 27, 28, 34, 39, 42, 56, 62, 71, 72
fuzzy controller, viii, 2, 3, 4, 7, 10, 11, 13, 15, 16, 17, 18, 27, 34, 42, 62, 71

fuzzy singleton model, ix, 42, 43, 45, 47, 49
fuzzy sliding mode control (FSMC), viii, 19, 20, 21, 23, 27, 31, 37, 39

G

genuinely Zeno, 82, 83, 86, 87, 89, 90, 93, 94, 96, 99, 102, 103, 105
g_i, 6, 23, 25, 45
graph, x, 117, 130, 133, 134, 135
guaranteed cost, 117, 120, 123, 128, 129, 130, 131, 132, 135, 150
guaranteed cost consensus, 117, 120, 123, 129, 130, 131, 132
guaranteed performance formation, x, 133, 134, 135, 137, 138, 142, 148
guaranteed performance function, x, 115, 133, 134, 135, 148

H

high-order, x, 115, 117, 130, 132, 133, 134, 135, 150
hybrid, 74, 75, 77, 78, 80, 83, 91, 112, 113, 132
hybrid arc, 78
hybrid signal, 78
hybrid system, 74, 75, 78, 80, 83, 91, 112, 113
hybrid time domain, 78

I

inequality, x, 80, 84, 86, 107, 115, 133
initial state, 81, 82, 84, 90, 92, 95, 96, 97, 100, 101, 103, 104, 105, 120, 121, 129, 141, 142, 147
input-output, viii, 19, 21, 22, 23, 24, 33, 37, 41, 44, 46
integration, ix, 62
interface, 60, 61, 69, 70, 72

internal model, ix, 42, 44, 54, 135
inverse model, ix, 42, 43, 44, 48, 49, 50, 51
inverse model control, 42, 43, 44, 48, 49, 50, 51
invertible, 49

K

king view, ix, 59, 61, 69, 70

L

Laplacian matrix, 119, 128, 129, 138, 147
least square estimation, 25
linear model, 5, 13, 15, 16, 39, 43, 56, 72
linear quadratic cost function, 119, 137
linear systems, 98, 112, 114, 132, 151
linear transformation, x, 115, 116, 117, 120, 121, 122, 133, 135, 139, 140
Lipschitz continuous on compact sets, 77, 79
local compression, 80, 84, 90, 95
local output equilibrium set, 84, 90, 91, 103
Lyapunov, 83, 101, 103, 112, 113, 124, 128, 135, 144, 146
Lyapunov function, 83, 144
Lyapunov-Krasovskii functional, 124

M

matrix, x, 6, 24, 26, 39, 46, 56, 67, 77, 95, 96, 99, 115, 117, 118, 119, 121, 123, 124, 125, 127, 128, 129, 130, 133, 135, 136, 137, 138, 139, 142, 143, 144, 145, 147, 148
matter, iv, 5
measurement, 26, 39, 56, 79
measurements, 43

media, vii, 21
membership, 7, 8, 9, 11, 23, 24, 25, 28, 46, 47, 63, 64, 66, 67
membership functions, 7, 9, 11, 23, 24, 25, 28, 29, 47, 63, 64, 66, 67
mobile robots, 39, 150
model prediction, 2, 3, 13
model predictor, viii, 1, 3, 4, 5, 10, 13, 15, 16, 39, 44, 49, 51, 54, 56, 72
models, 22, 30, 33, 36, 37, 51
motion control, 30, 38, 40
motor system, 50
MPC, 17, 38, 57
multi-agent system, viii, x, 111, 113, 114, 115, 116, 117, 118, 120, 122, 128, 129, 130, 131, 132, 133, 134, 135, 137, 138, 139, 140, 148, 149, 150
multiplier, 24

N

NetCon system, viii, 41, 42, 43, 45, 50, 52, 54
networked control, 1, iii, vii, viii, ix, 1, 2, 3, 4, 17, 18, 19, 20, 29, 34, 38, 41, 42, 49, 54, 55, 56, 57, 59, 60, 71, 112, 153
networked control system, 1, iii, vii, viii, 2, 3, 4, 17, 18, 19, 20, 38, 42, 55, 56, 57, 60, 112
networked predictive fuzzy control, v, vii, 3, 4, 7, 10, 13, 18, 39, 56, 72
networked tracking control, v, vii, viii, 19, 20, 21, 31, 39, 41, 43, 56
NFPC, 13, 15
nonlinear dynamics, 33, 135
nonlinear systems, 39, 112, 114, 131
norm-inducing, 24

O

online information, ix, 73
OPC interface, 60, 61
optimization, ix, 23, 24, 59

P

PID controller, 15, 31, 32, 34, 52, 60, 61, 63, 68, 71
PID tuning, 60
PLC, ix, 59, 60, 69, 71
point-wise, 24, 25
prediction, vii, 2, 3, 13, 21, 26, 27, 31, 36, 38, 42, 43, 44, 51, 54, 55, 56
predictive control, 3, 17, 18, 20, 38, 39, 40, 43, 55, 56
proceedings, 18, 38, 39, 56, 57, 72, 111, 112, 113, 131, 132, 150

Q

quantization, 56, 111

R

random delay, viii, 2, 15, 16
random time delay, 14, 15, 54
real-time, 7, 10, 13, 16, 18, 43, 44, 69, 71, 72, 74, 111
real-time control, 7, 71, 72
relative triggering condition, ix, 73, 76, 80, 93, 94, 98, 99, 100, 101
remote supervisory, 60
researchers, 3, 43, 116, 134
resources, ix, 73, 74, 79
response, 6, 13, 32, 34, 52, 62, 69, 71, 98
response time, 71

robust guaranteed performance, vi, x, 115, 116, 117, 133, 134, 135, 137, 138, 142, 143, 146, 148
robustly stable, 101, 141, 142
rules, 7, 8, 9, 10, 11, 28, 64, 65, 66

S

scaling, 10, 28, 66, 68, 71
Schur Complement, 123, 127, 132, 142, 151
servo control system, 4, 13, 29, 30, 43, 44
signals, ix, 30, 44, 51, 54, 69, 74, 75, 76, 93, 110
simplified, 5, 7, 18, 34, 36, 37, 51, 64
simplified fuzzy inference, 64
solution, vii, 3, 24, 26, 75, 76, 78, 81, 82, 84, 85, 86, 87, 89, 90, 91, 92, 93, 94, 96, 98, 99, 101, 102, 103, 105, 106, 107
structure, ix, 3, 4, 16, 20, 21, 22, 37, 42, 43, 44, 45, 50, 54, 60, 63, 71, 118, 141

T

technology, viii, ix, 2, 19, 22, 23, 37, 42, 44
temperature, ix, 59, 60, 61, 68, 69, 71
temperature control system, 60, 61, 68, 71
time delay, vii, viii, 2, 14, 15, 17, 18, 20, 21, 29, 37, 38, 39, 41, 42, 50, 54, 55, 57, 117, 121, 129, 131, 132, 150
time signals, 74
time-varying delays, x, 115, 116
topology, x, 116, 117, 118, 119, 128, 129, 130, 133, 135, 137, 138, 147, 148, 150
transformation, x, 115, 116, 117, 120, 121, 122, 133, 135, 139, 140

transmission, ix, 5, 12, 17, 28, 42, 51, 73, 74, 76, 82, 83, 90, 91, 108, 109, 116, 135

triggering condition, ix, 73, 74, 75, 76, 79, 80, 82, 83, 86, 87, 88, 89, 91, 92, 93, 94, 95, 96, 98, 99, 100, 101, 102, 103, 104, 105, 106, 107, 108, 109, 110

T-S fuzzy model, viii, 23, 41, 43, 45, 55, 56

U

uncertain systems, 116, 120, 139

uncertain topologies, vi, 117, 133, 134, 135, 148

uncertainties, x, 114, 115, 127, 135, 137, 150

uncertainty, vi, ix, 42, 44, 115, 117, 118, 119, 120, 121, 130, 145

V

vector, 24, 46, 49, 51, 77, 78, 79, 117, 136, 137, 141

velocity, 5, 33, 129, 136, 137, 146, 147

Z

Zeno behavior, v, ix, 73, 74, 75, 76, 81, 82, 83, 84, 86, 90, 91, 95, 96, 99, 101, 102, 103, 104, 105, 106, 107, 110, 113

Zeno equilibrium, 75, 82, 84, 86, 87, 88, 89, 90, 91, 92, 93, 94, 95, 101, 102, 103, 106, 110

Zeno instant, 81, 85, 86, 87, 99

Zeno-free, x, 74, 76, 81, 93, 96, 98, 107, 108, 109, 111